Arduino App Bluetooth Robotics

Dr Kesorn Pechrach Weaver

Arduino App Bluetooth Robotics

Dr Kesorn Pechrach Weaver

Pechrach Publishing
United Kingdom

Arduino App Bluetooth Robotics

By Dr Kesorn Pechrach Weaver

ISBN 978-1-912957-06-4

PECHRACH PUBLISHING

7 Boundary Road, Bishops Stortford, Hertfordshire, CM23 5LE, England, United Kingdom. Tel: (+44) 1279 508020, +44 (0) 7443426937

Published 2019 by Pechrach Publishing

This book is dedicated to my robotics students

Message from the Authour

After I conducted a robotics class for my young students in the primary and secondary school year 6 to year 10, age 11 -17 years old.

I realise that their brain, their creative and their imaginary are very powerful. Thus, they are ready to create their own mobile App and control their robots using Bluetooth and Arduino. Therefore, this book is the next part of the Educational Robotics.

Dr. Kesorn P. Weaver

21st May 2019

Bishops Stortford, UK

Acknowledgments

I would like to thank Dr Paul M. Weaver and Neran J.P. Weaver for their support both mentality and physicality during preparing this book.

Many thanks to my students from the Computer Science and Robotics Class; Thomas Wilkinson, Ruby Luck, Joshua Williams, Jorja Cox, Harry McCrae, Eleanor Chichlowska, Benjamin de Vos, Aaron Walker, Alex Rodriguez, Deborah Toms, Ellie Simeonova, Felix Baker, Hattie More, Naomi Williams, Ollie Hall, Paige Tritton, Tom Gibbs, Ander Miera, Ben Cooper, Florence Chichlowska, Hannah Walker, Freddy Brazier, Jack Cooper and Scarlett Murray. They are brilliant and provide me with a new vision and dimension of thinking.

A special thanks to M Emlyn Humphries, Lisa Nichols, Rachel Kelly, J.P. Darby and Rung Ratpinyyotip, for all their support.

Thanks to my family in Thailand, my friends and my family in the United Kingdom for believing in me.

Introduction

This is a small book and show clear description step by step on how to build App, to program, connect circuits and run the simulation test before building the real hardware robots.

The students should be able to follow step by step easily. There are five chapters in this book. Thus, the easiest way to do is start from the chapter 1. However, this whole book could follow and finish within a couple hours.

Table of Contents

Acknowledgement

CHAPTER 1

Build our Own App

Create Your Own App

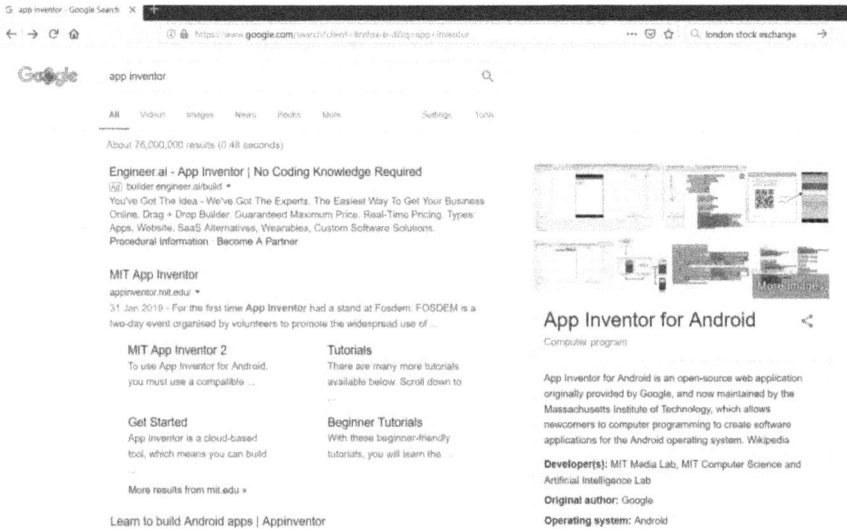

Search google the word "App inventor" and choose the MIT Inventor.

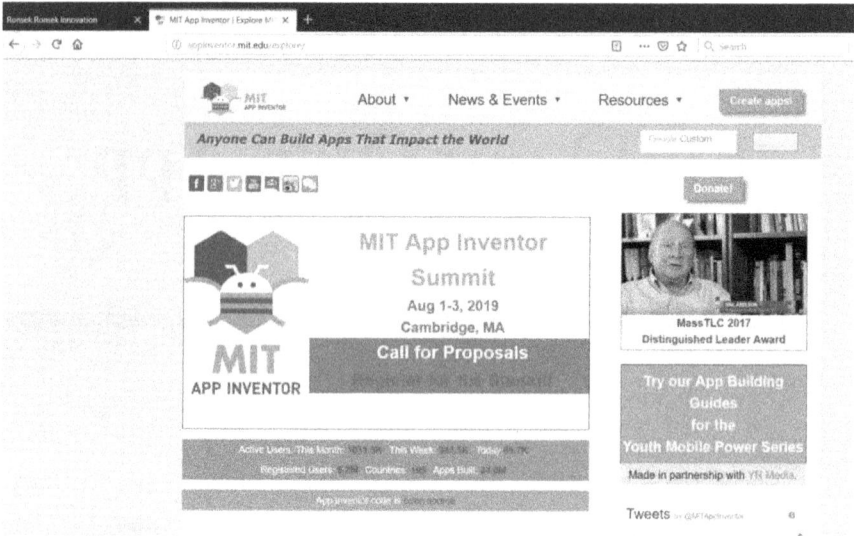

That is the webpage of appinventor.mit.edu

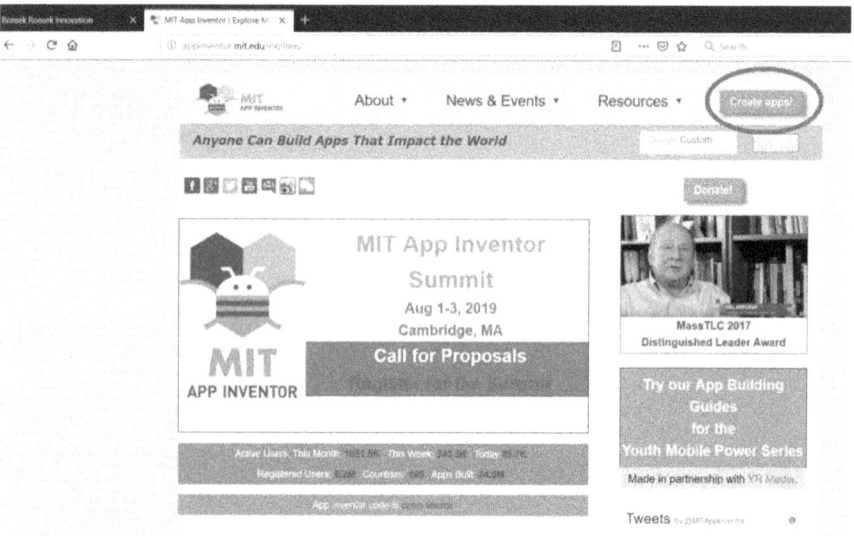

Click on the bar "Create Apps!", there is a page to sign in or create an account.

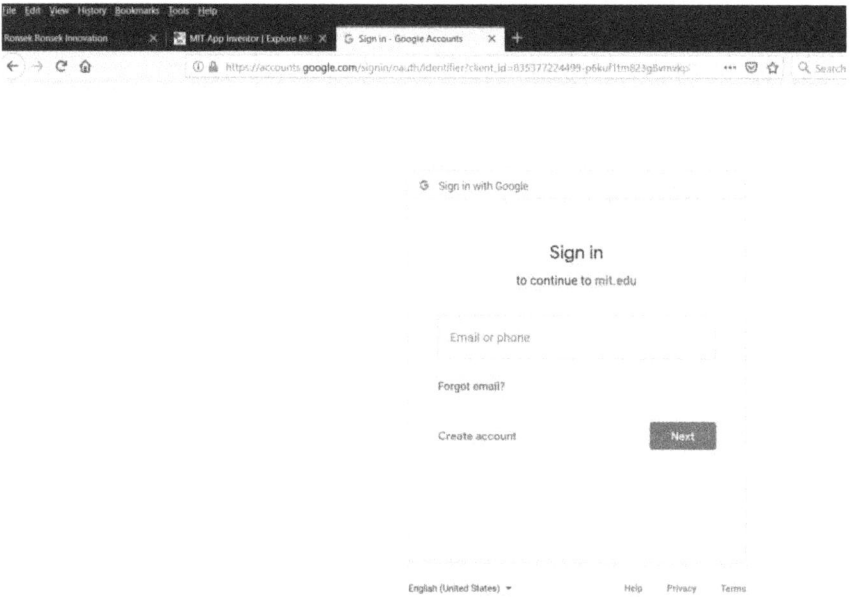

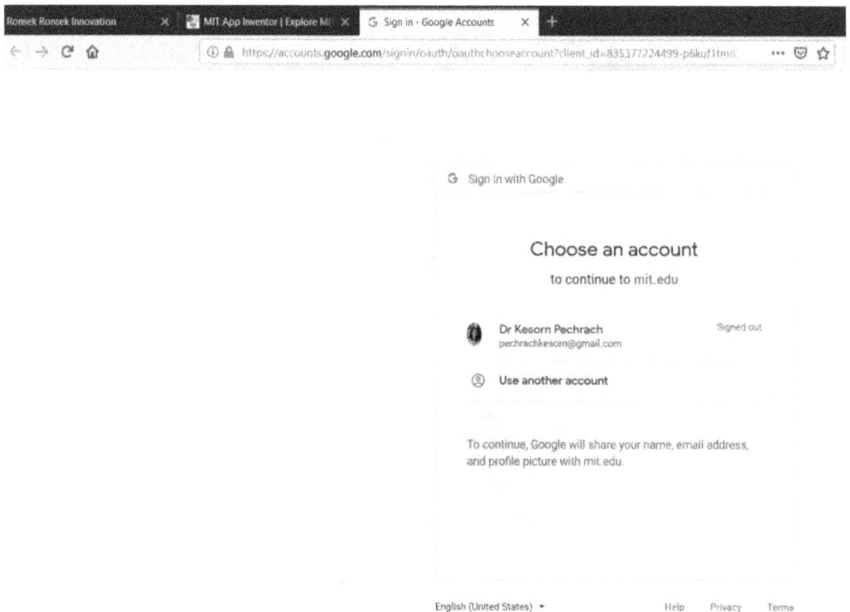

After we click to sign in, the screen would show as follows:

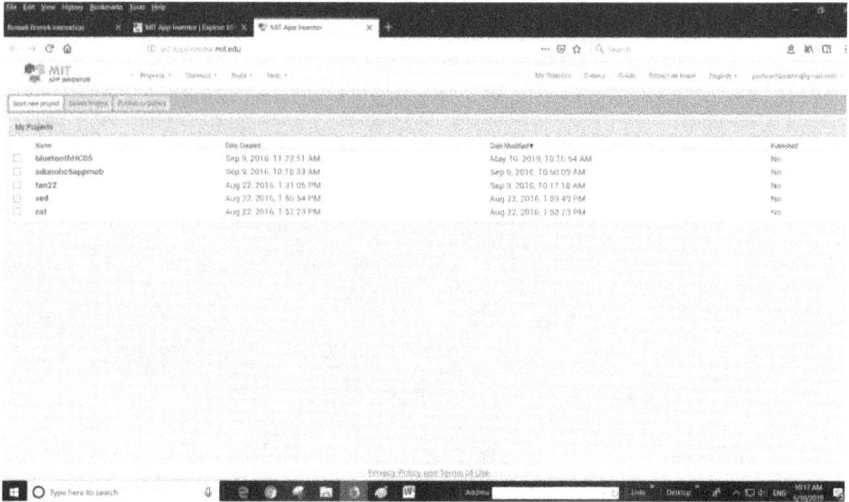

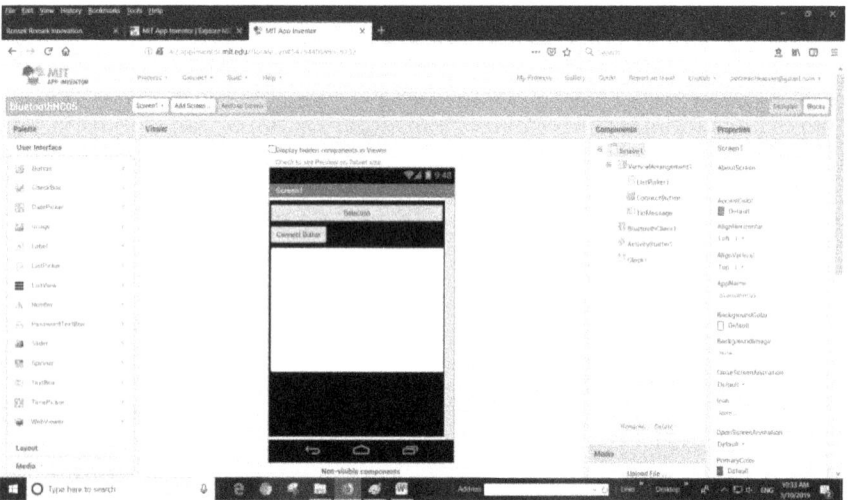

Now we are ready to start our new project.

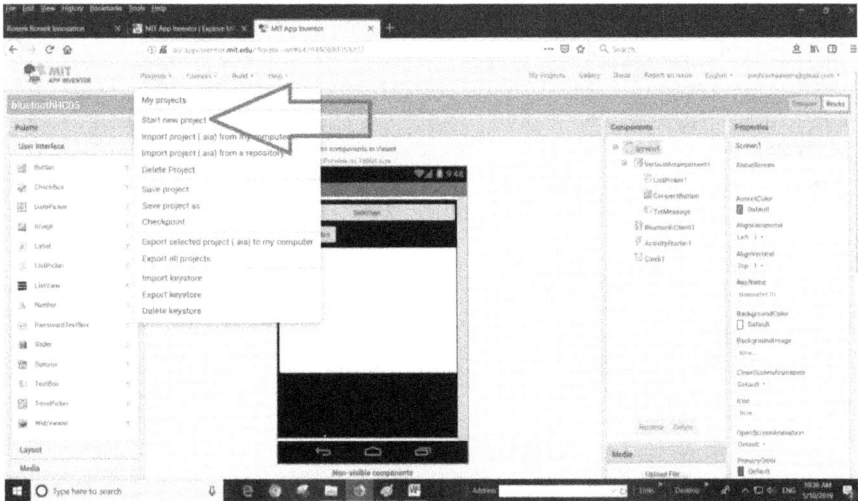

Go to "Project" on the top bar and choose " Start new project"

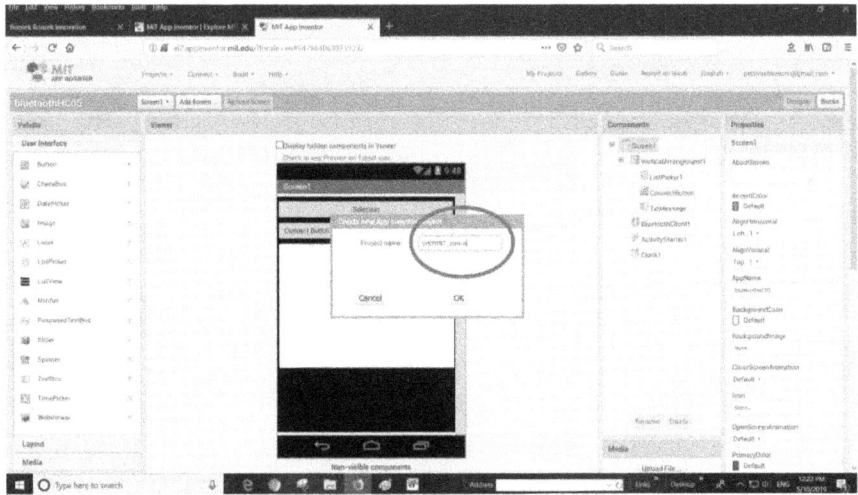

We give a name for our new project and click "OK". Thus, we will have a new window show our new project name.

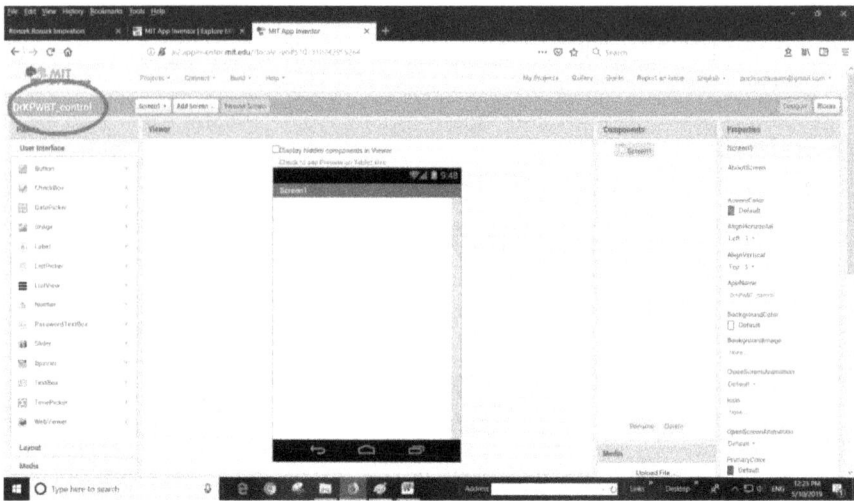

Next, we choose the ListPicker, this tells which the device we would use to connect.

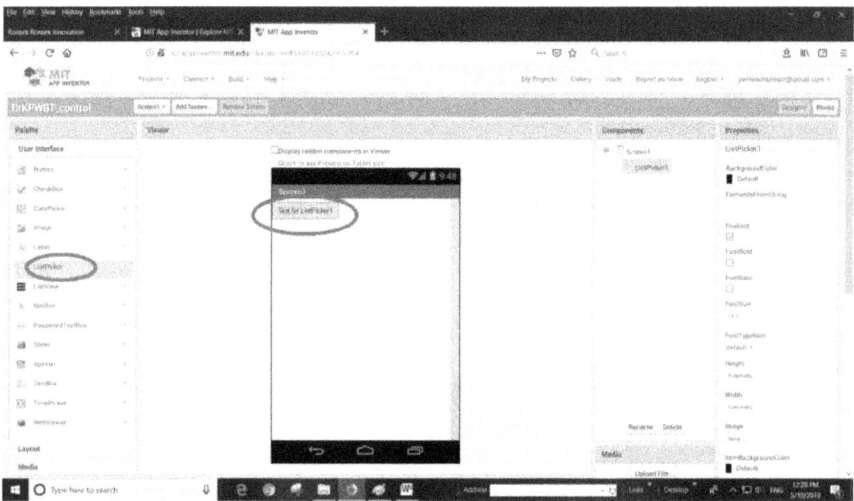

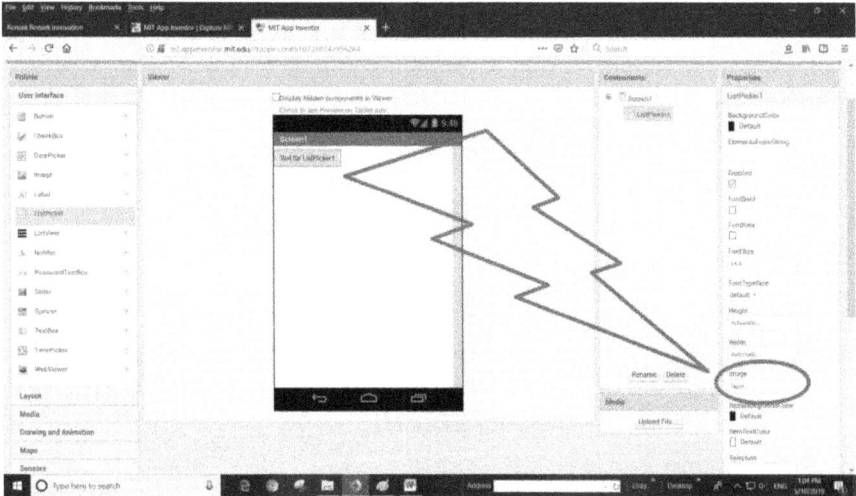

To add pictures to the ListPicker, click at "Image" on the side bar and click upload the picture from our computer. Choose a picture and click "OK"

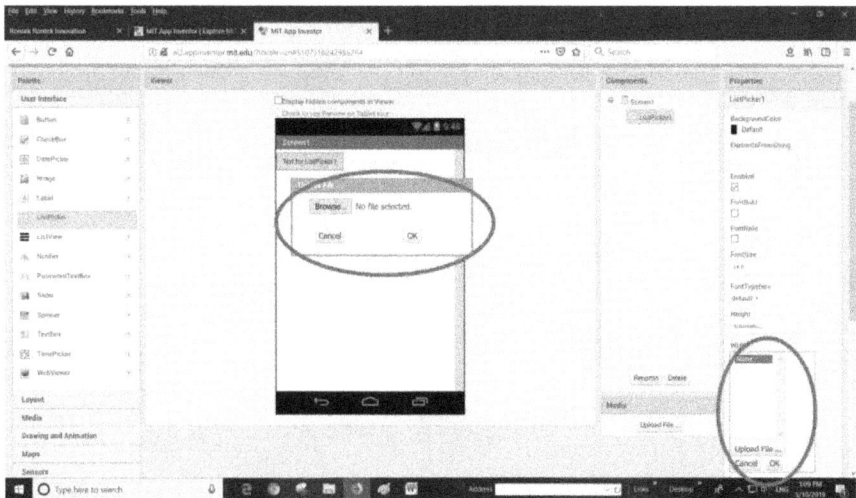

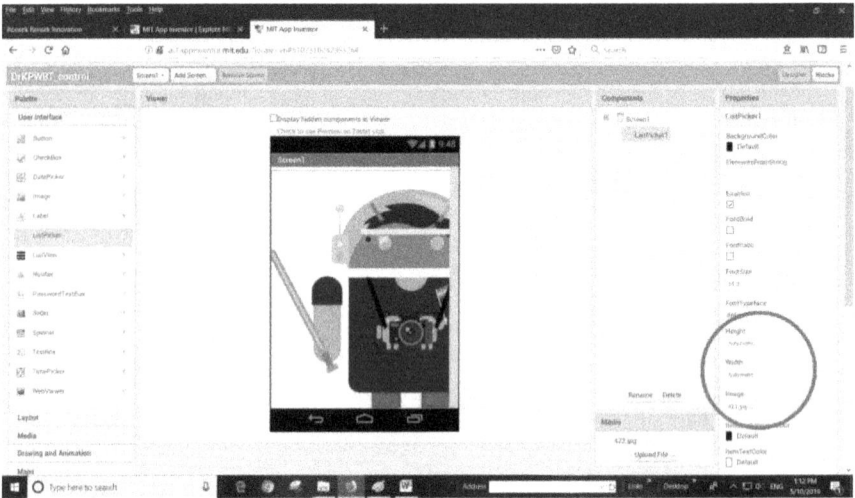

Unfortunately, the picture is too big. Therefore, we need to adjust the picture size: Height and Width. This would make the size smaller than before.

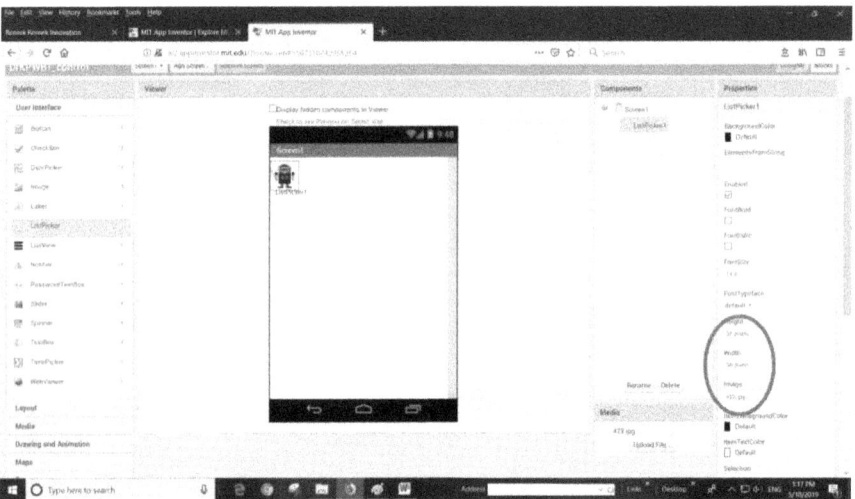

We want to get rib of the text in the middle, we can do delete the text in the text block.

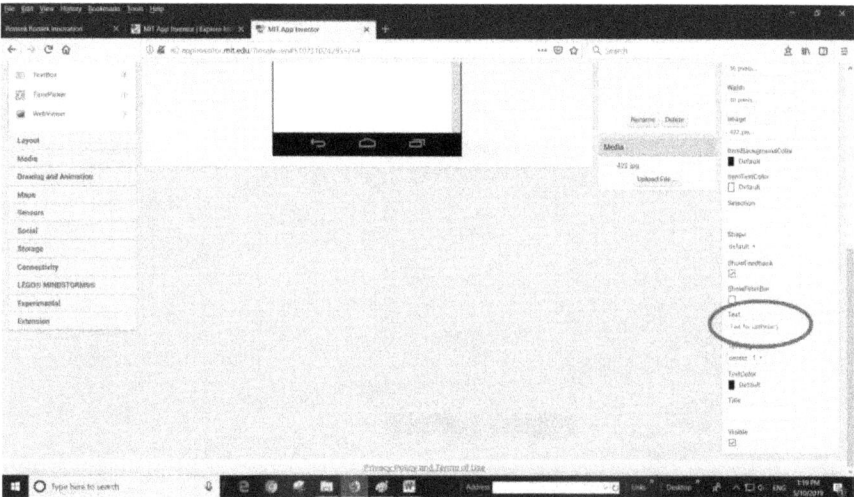

Next, we will add the button, drag and drop in our screen. Then, go to get a picture from the Image sidebar and change the size.

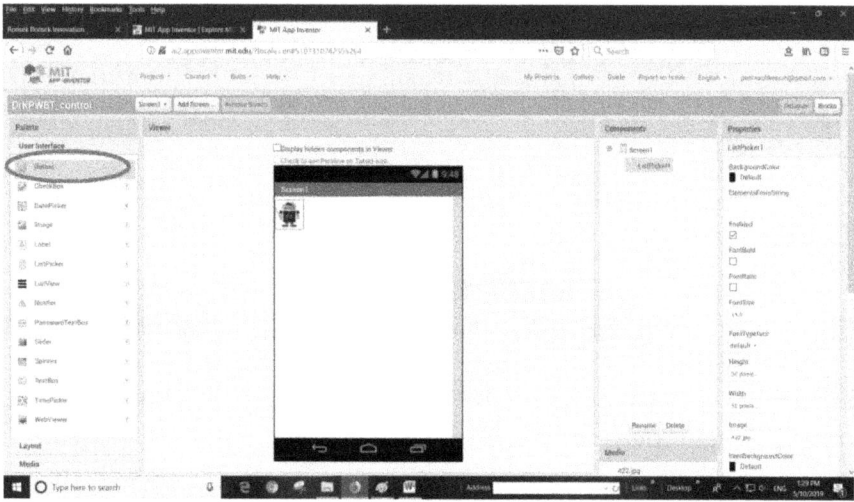

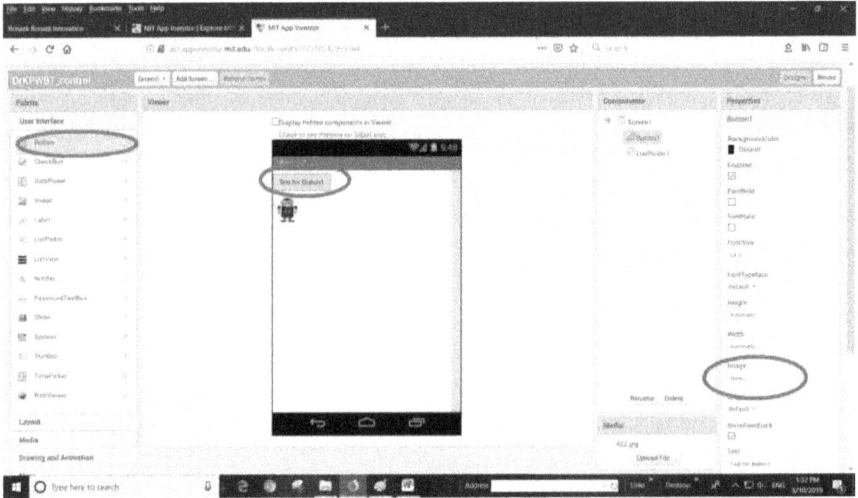

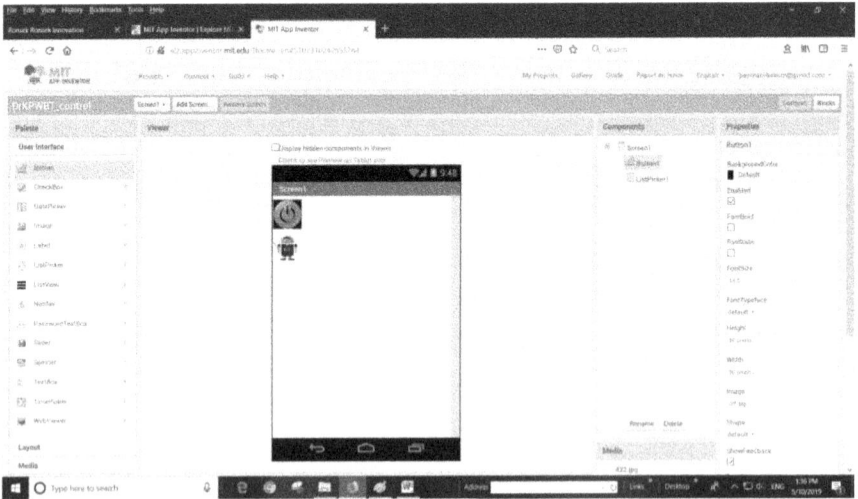

We will have pictures as below. However, if we want they are on the same line. We will go to Layer.

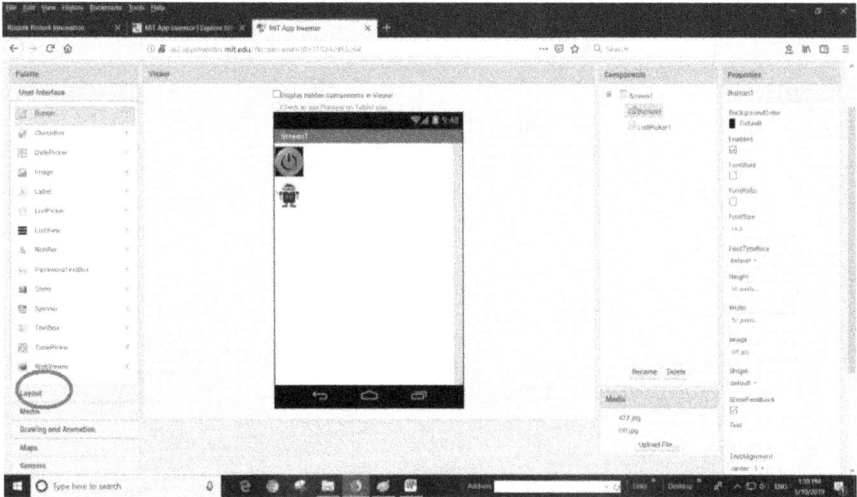

In the Layout, choose "Horizontal Arrangement" and put it on the screen. We will see it as a square box.

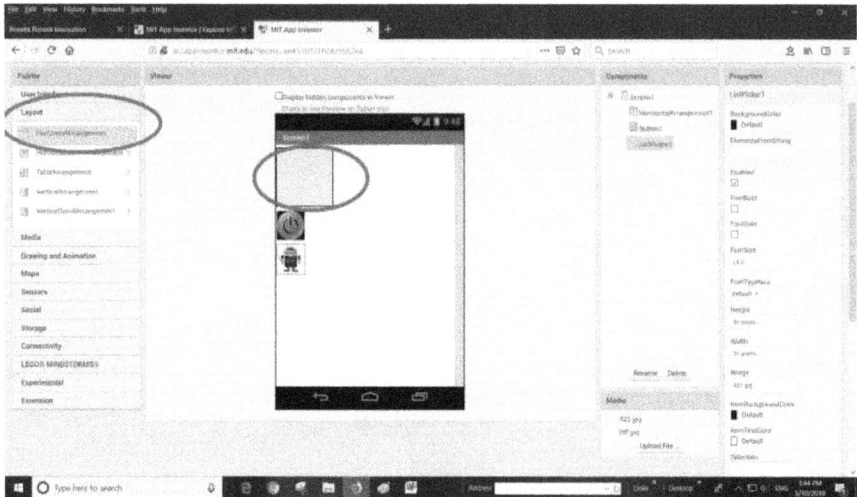

Next, move both Bluetooth and switch button in the box.

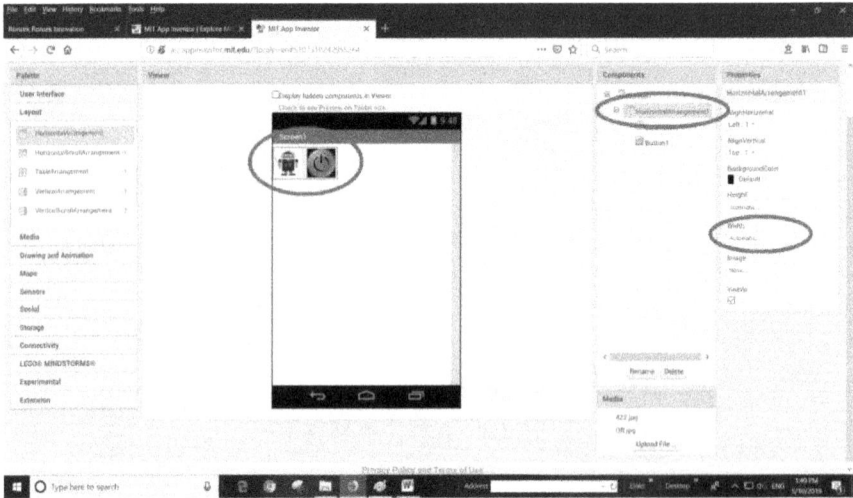

Then, adjust the size of the box . At the width pixel 310 is the length.

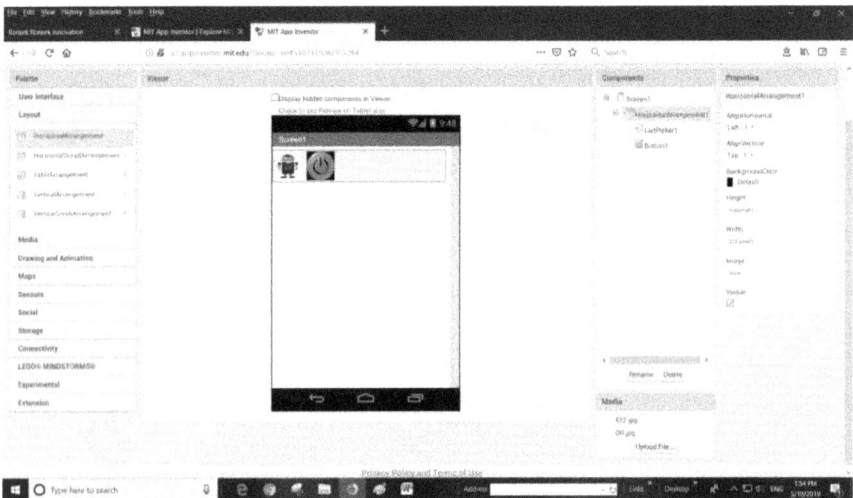

We want to have a gap between them. The "HorizonScrollArrangement" was chose to drop between them and send the width to 175 pixel.

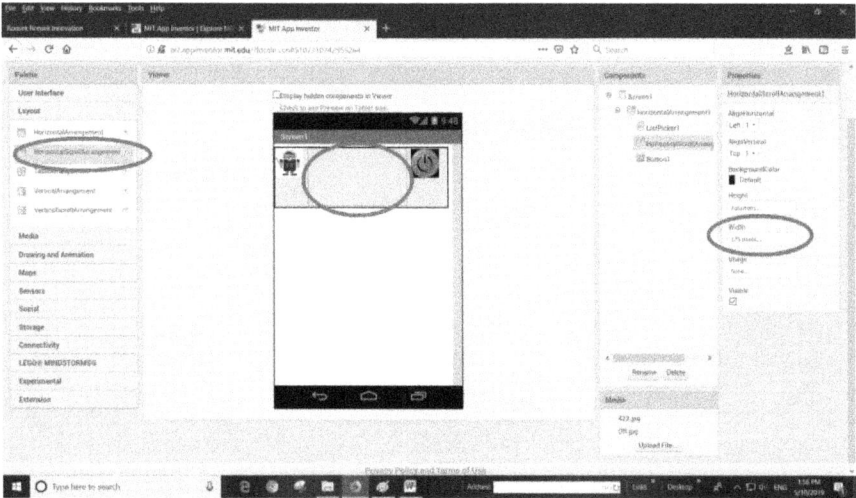

The background colour can change as we like.

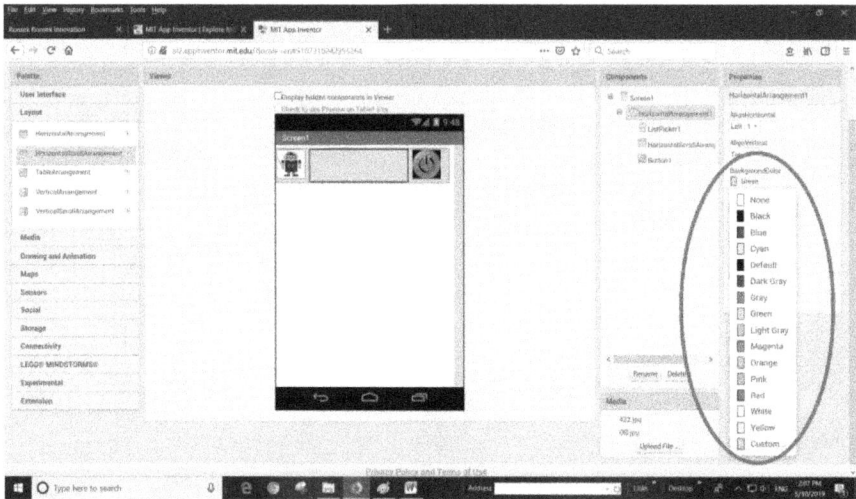

Next, we would like to have more button, we choose the "HorizontalArranagement" and place below.

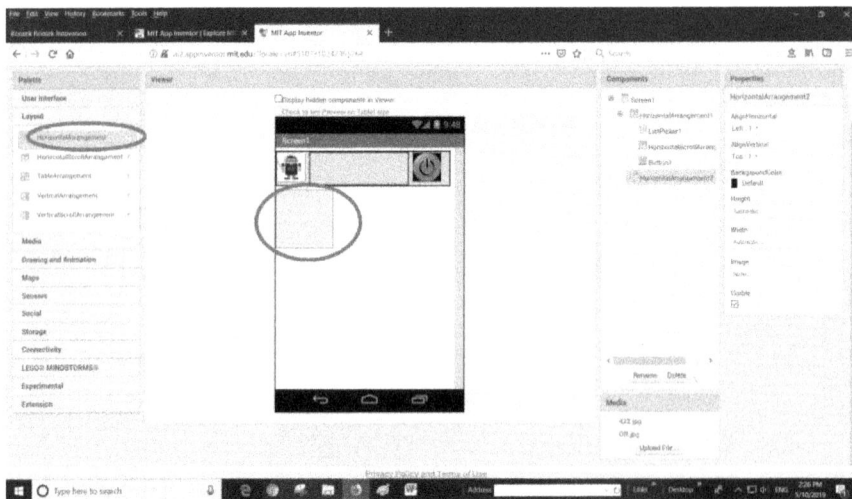

We set the Width to 315 pixel.

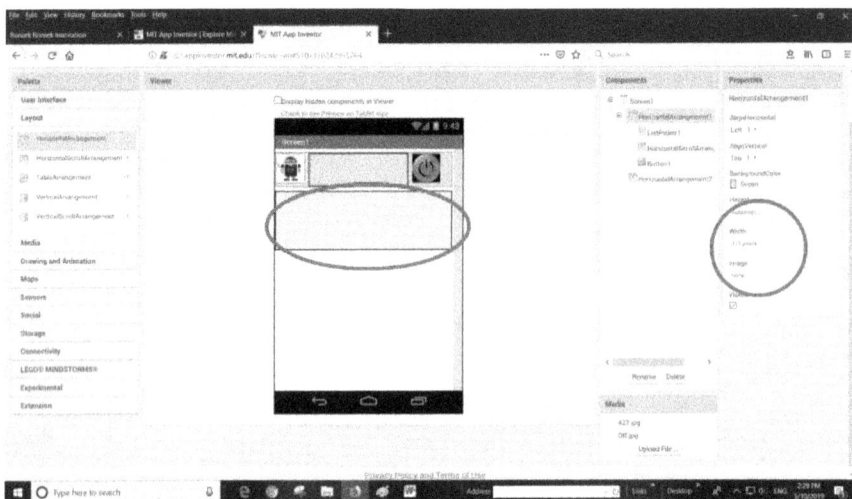

Add more button from the "User Interface"

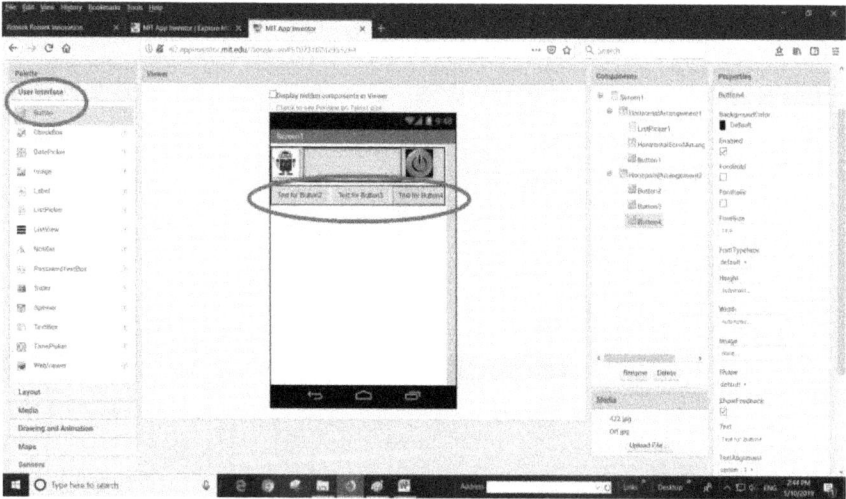

Add pictures, delete the text in the text block and adjust the size.

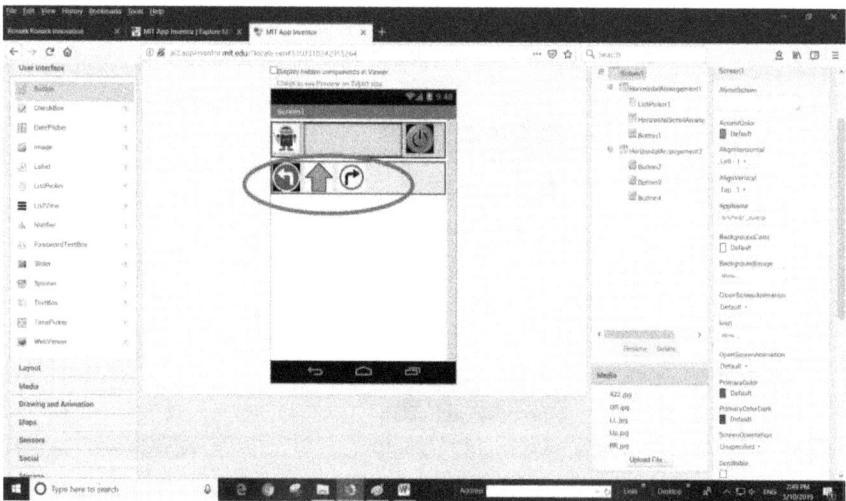

We want to have gaps between them, use the HorizontalScrollArrangement in the "Layout".

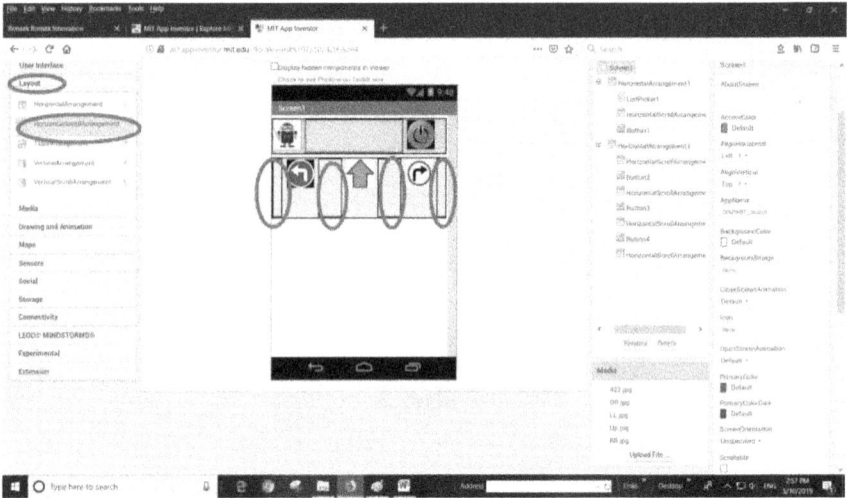

Next, we will repeat those steps with two more rows.

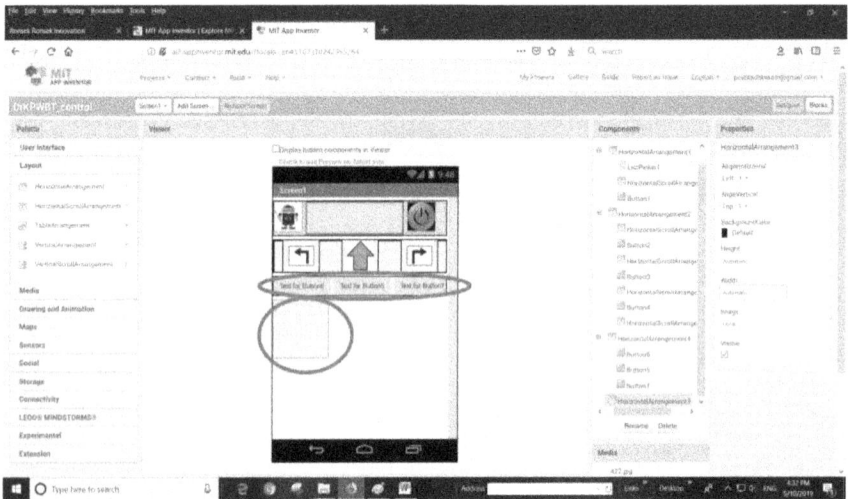

Add pictures in the second row and use the "HorizontalArranagement" in the Layout in front of the first button.

Arduino App Bluetooth Robotics

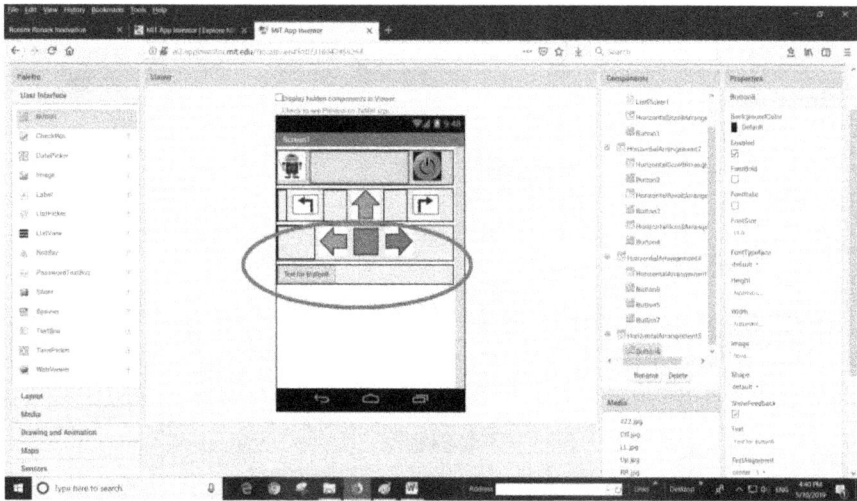

We have all buttons that we need to control the directions.

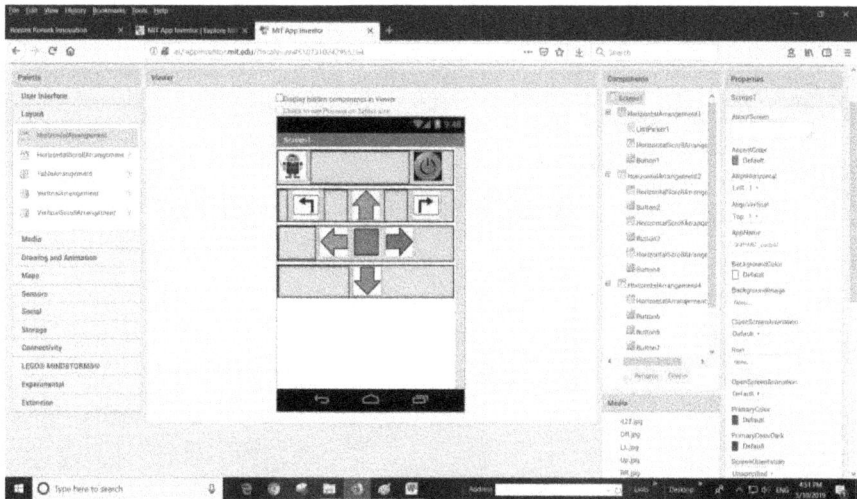

Next, we are going to code for the Bluetooth. At the connectivity, choose the "BluetoothClient", drag and place in the space.

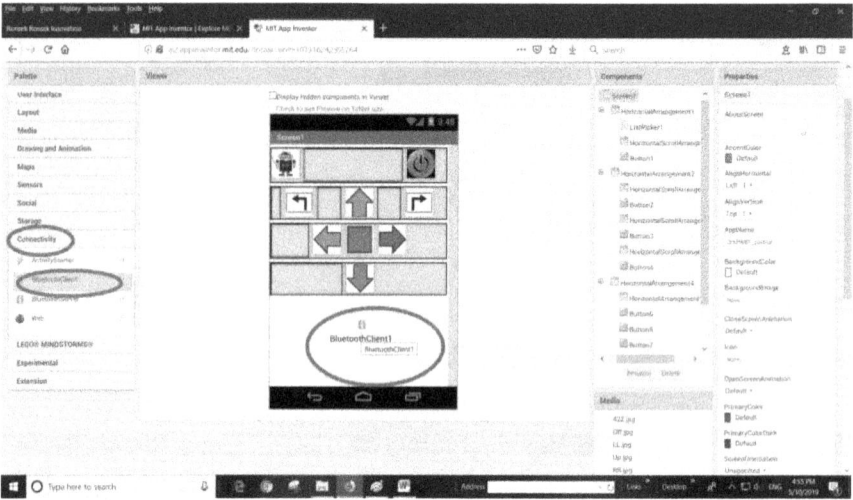

Thus, the BluetoothClient would be in the Non-visible components.

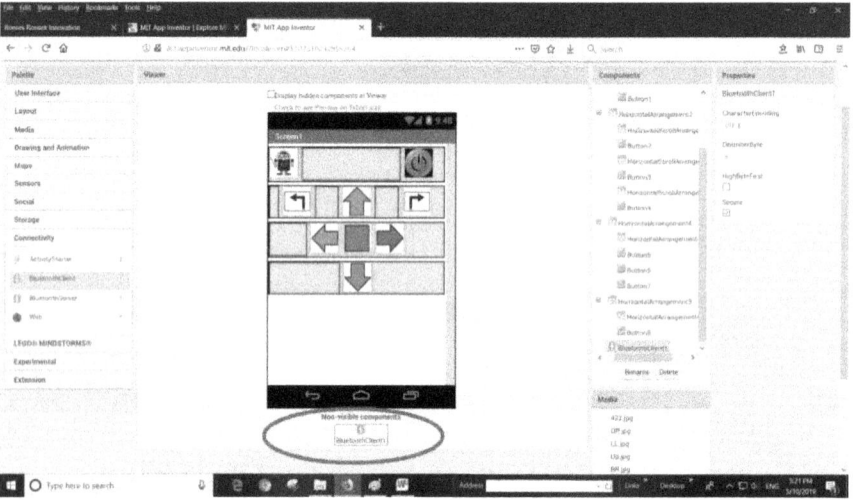

CHAPTER 2

Code our App

A. Connect to Bluetooth

We have all buttons that we need to control the directions.

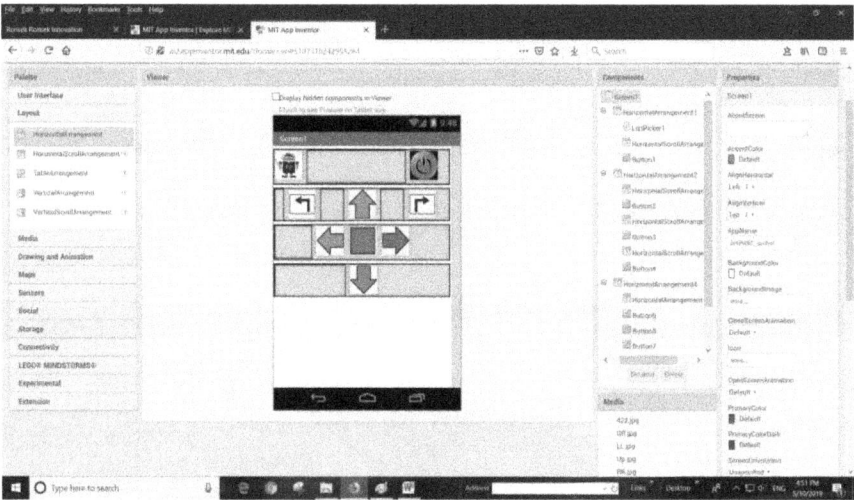

Next, we are going to code for the Bluetooth. At the connectivity, choose the "BluetoothClient", drag and place in the space.

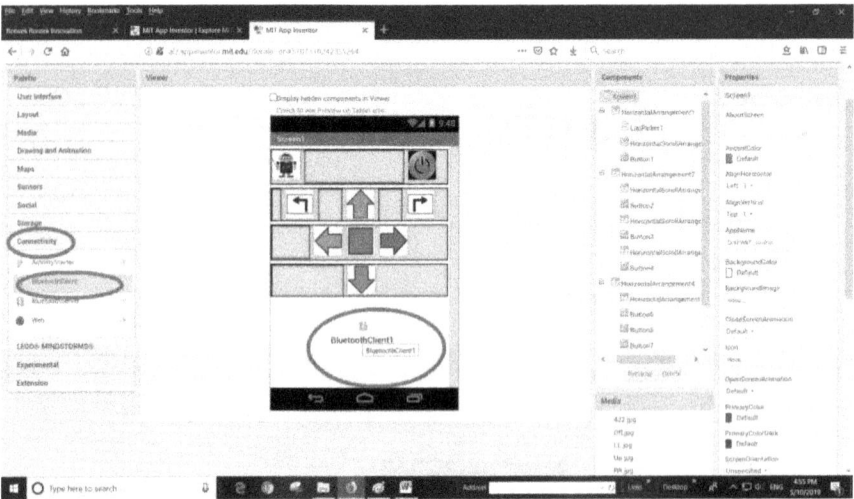

Thus, the BluetoothClient would be in the Non-visible components.

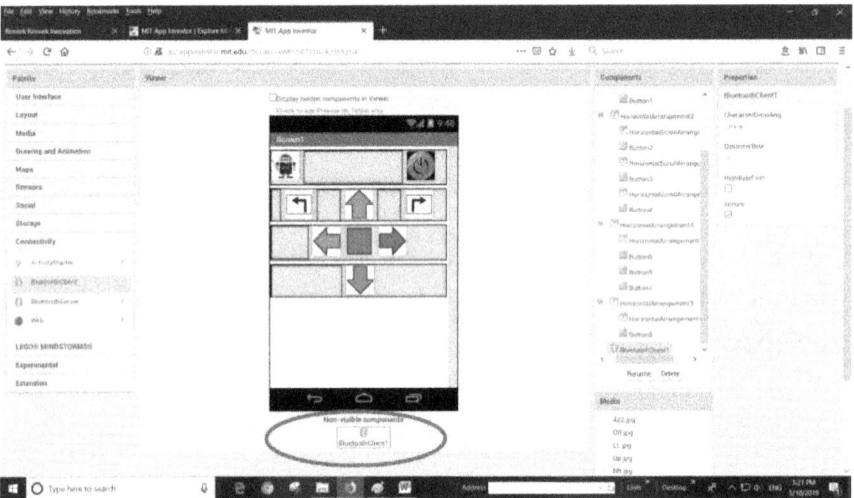

Next, we will start coding, click on the top bar right hand side "Blocks".

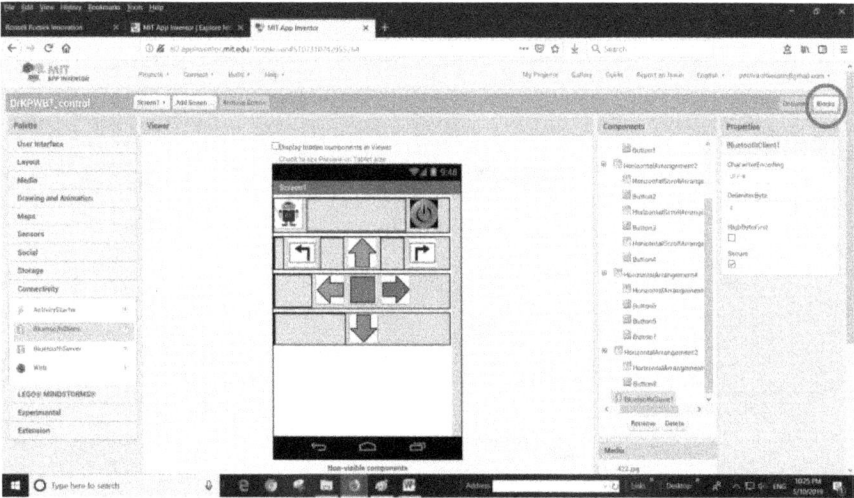

The new screen would be as below.

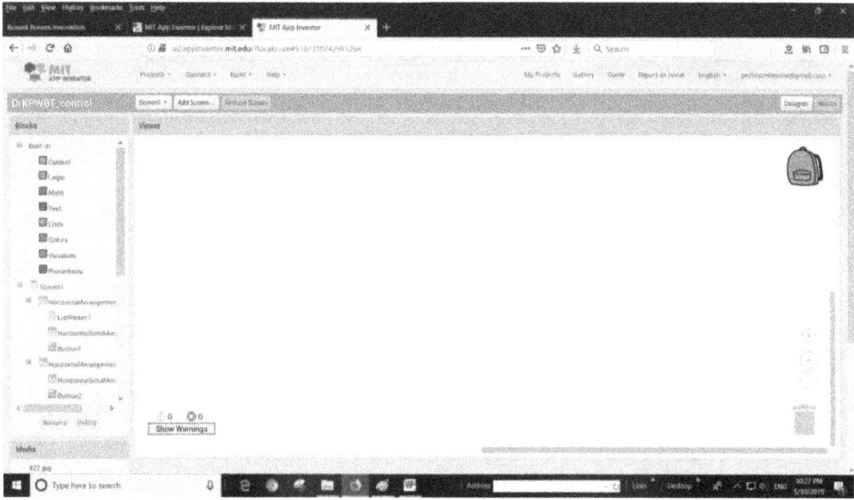

We can toggle between two screens: Designer and Block. The first command is the Bluetooth, which we use the "ListPicker1" on the Designer Screen.

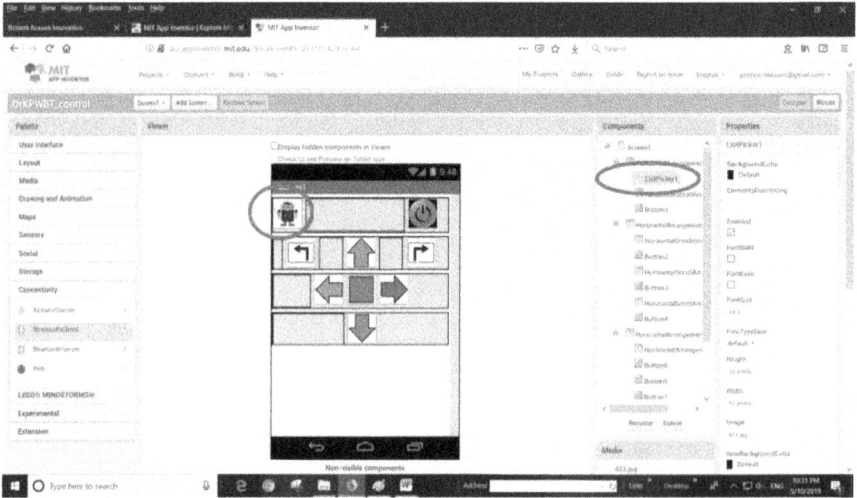

On the Blocks Screen, the "ListPicker1" is on the sidebar, drag and drop on the space on the right hand.

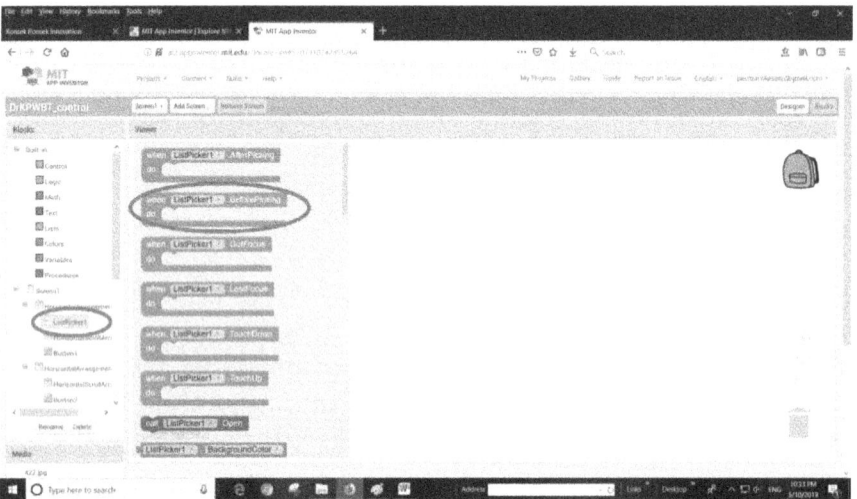

Arduino App Bluetooth Robotics

Next, we choose the do function to connect Bluetooth.

Move the green command and connect to the Yellow LisPicker1.

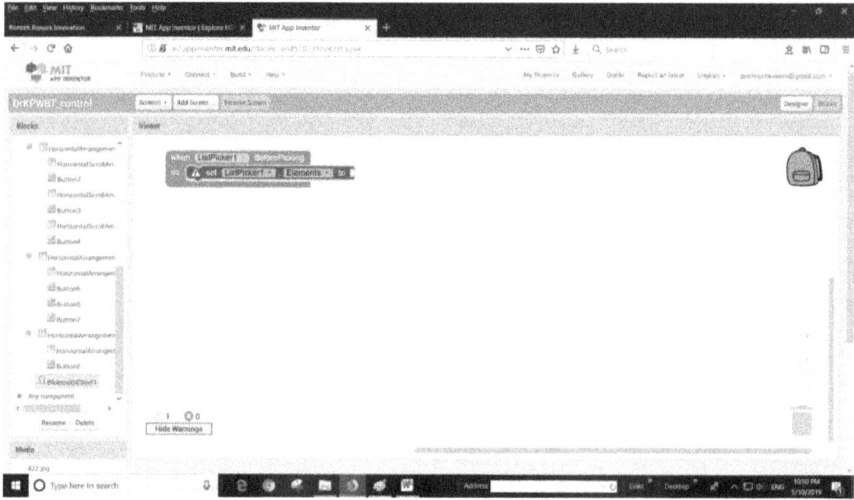

The BluetoothClient would allow to connect to all available Bluetooth.

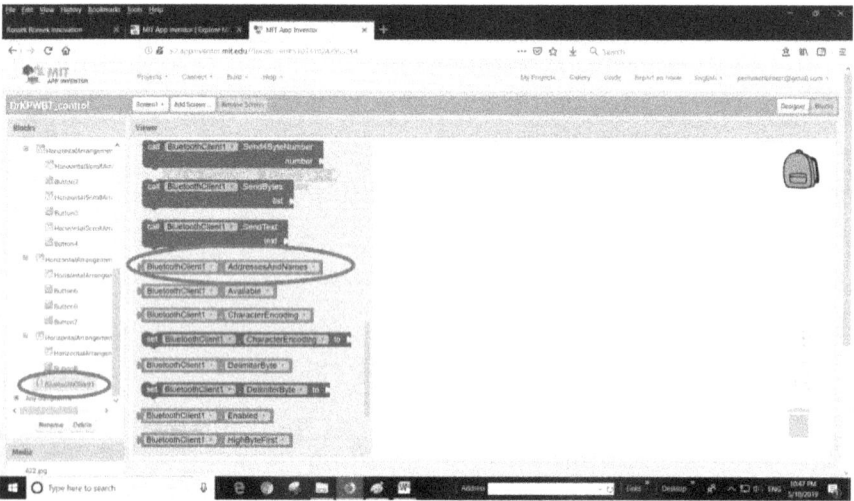

The command before picking.

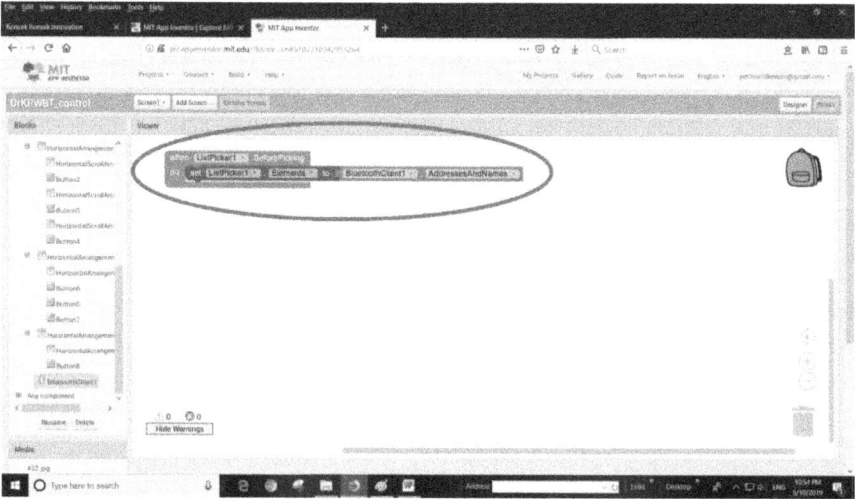

The next command is after picking, it will choose to connect Bluetooth.

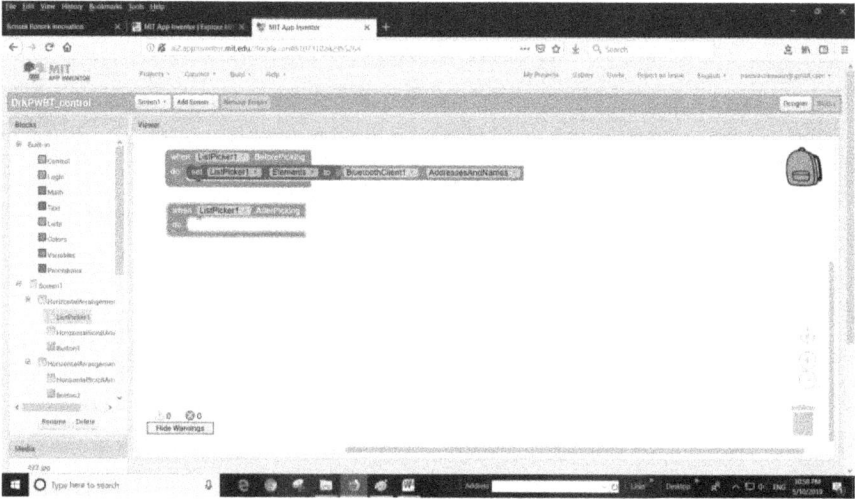

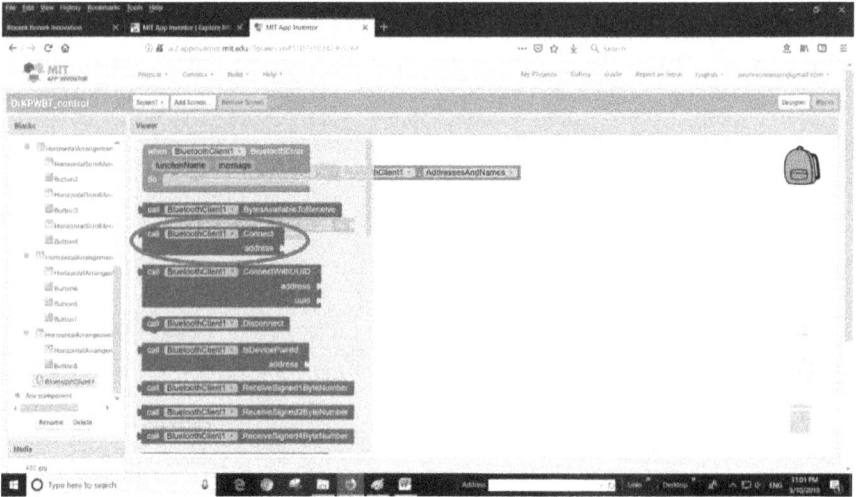

Connect ListPicker to Bluetooth.

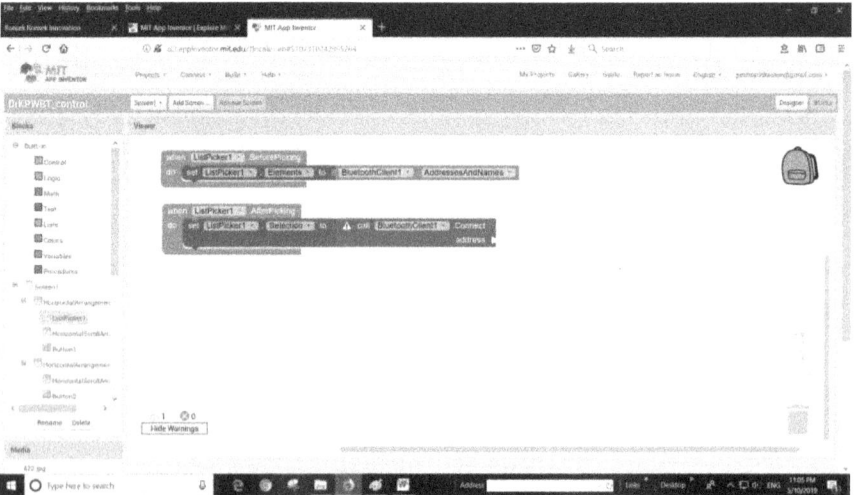

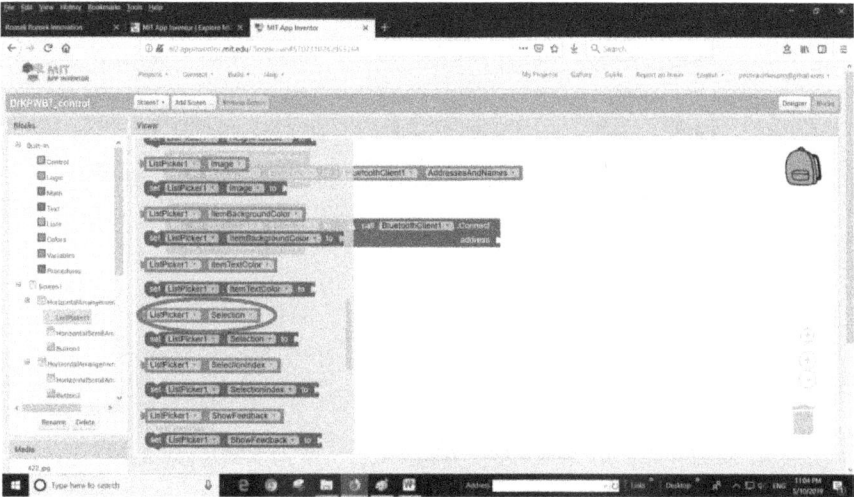

The command for connecting Bluetooth is done.

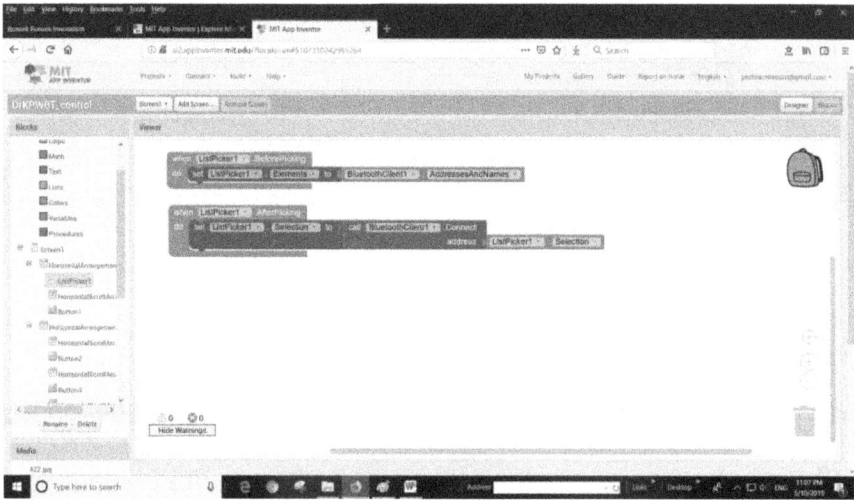

Next is the Button 1, which it would be used to turn it off.

B. Turn Off Screen: Button 1

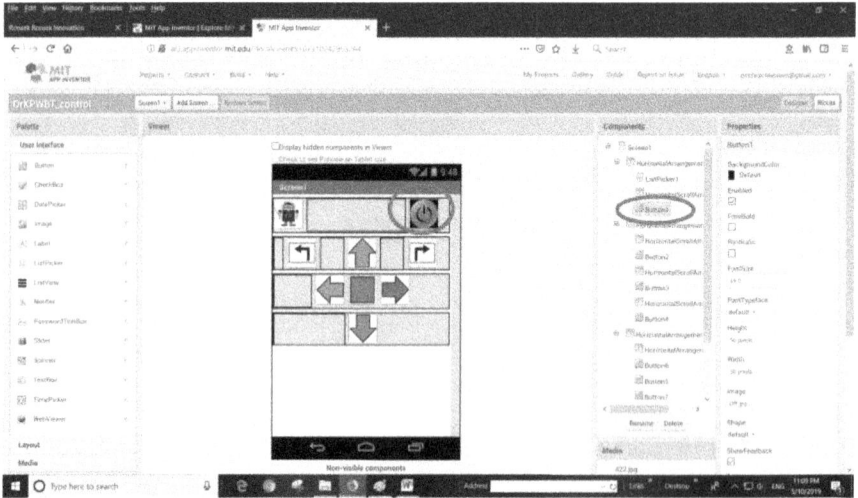

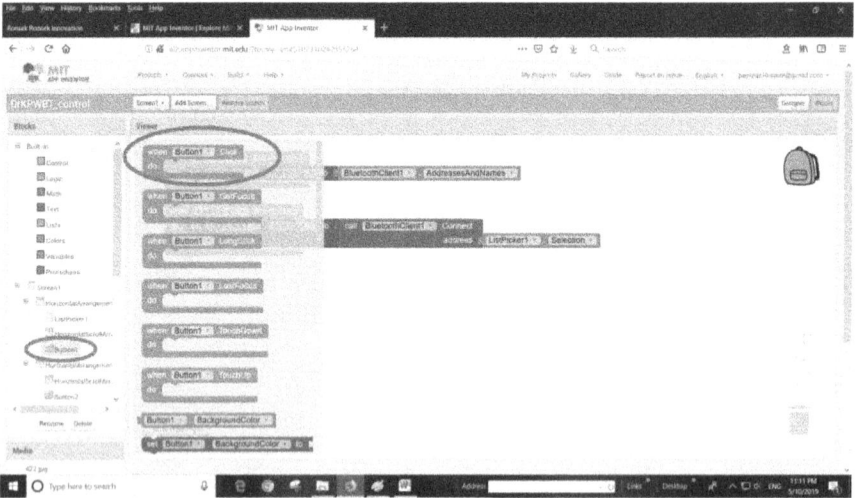

Choose the " command ", when the Button 1 click".

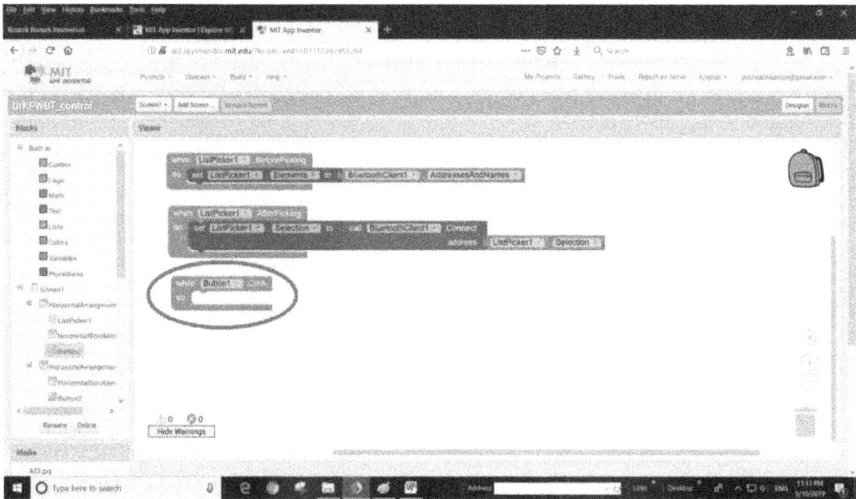

The Bottle 1 is for closing the screen.

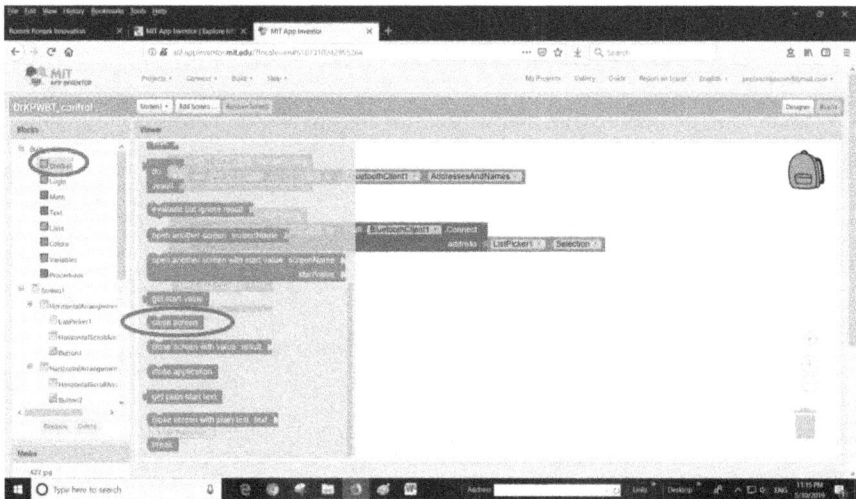

The command we want is the " Close the screen".

We have finish the first row; connect to Bluetooth and Close the screen

CHAPTER 3

Code and Button

A. Left Anticlockwise Circular Button

The Button 2 is the "Turn Left Anticlockwise Circular" on the Blocks screen

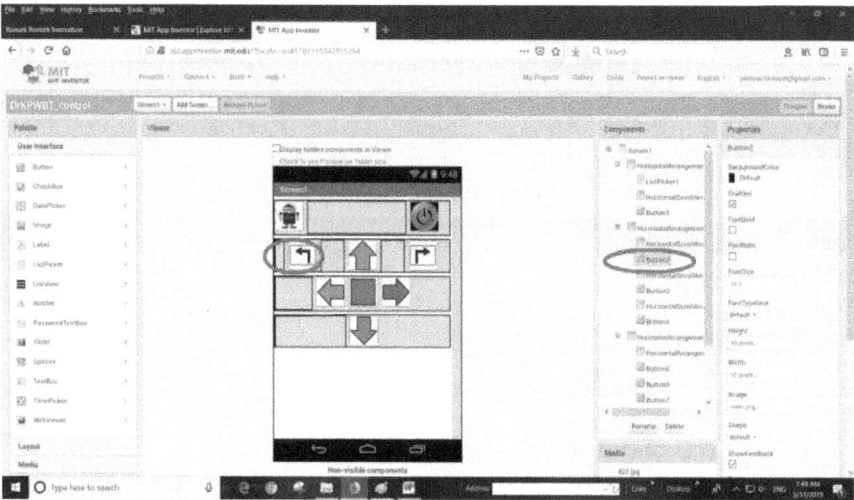

Thus, we will work on the Button 2 in the Designer screen.

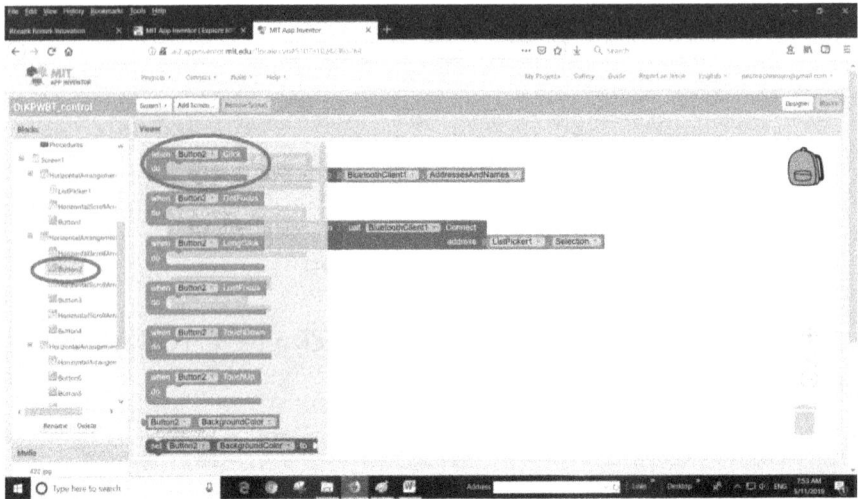

We choose "When Button 2 click" from the Button 2.

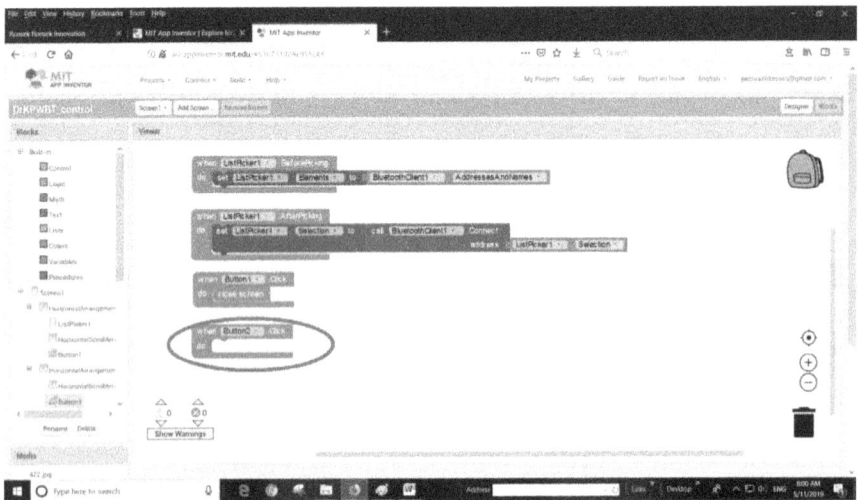

Arduino App Bluetooth Robotics

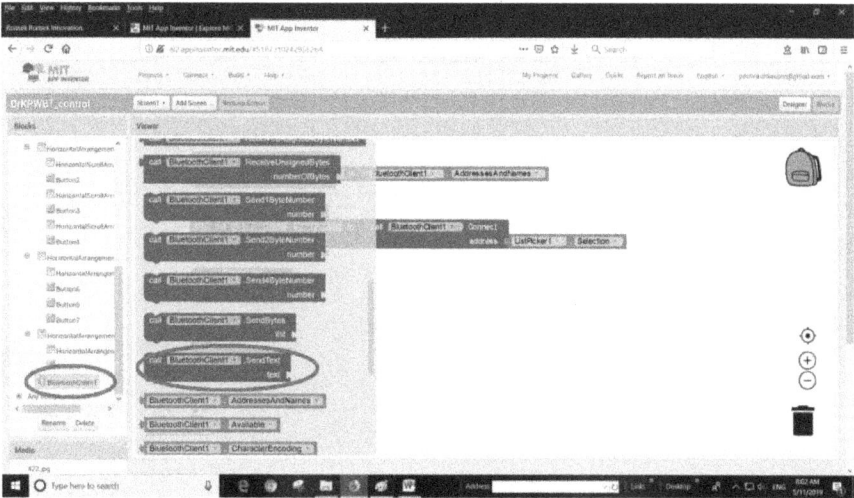

We need it to send Text from BluetoothClients.

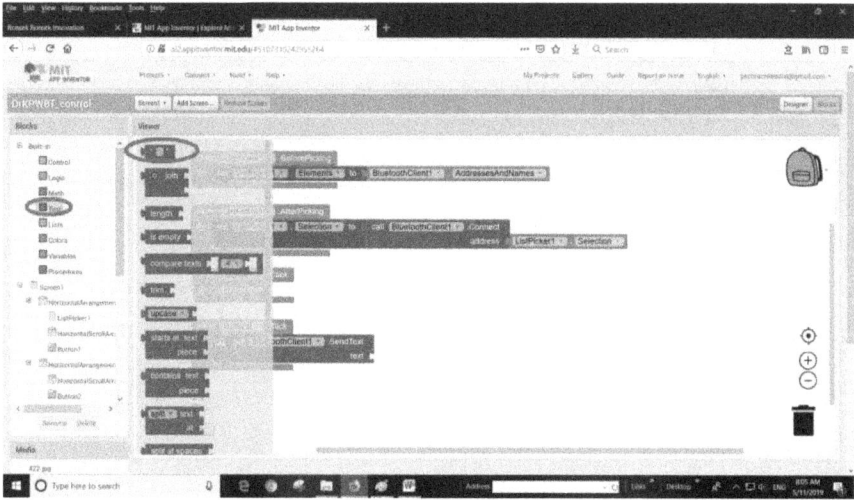

We use the letter "L" for left circular anticlockwise movement.

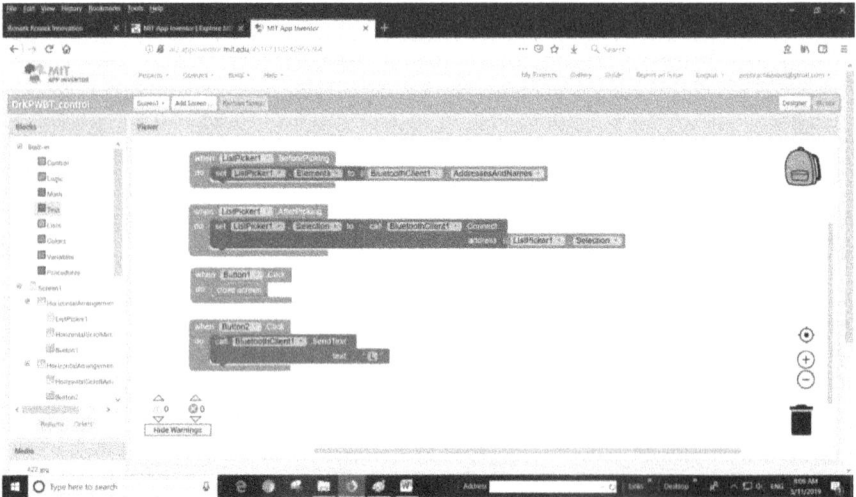

B. UP Button

Button 3 is the function to go up or move forward.

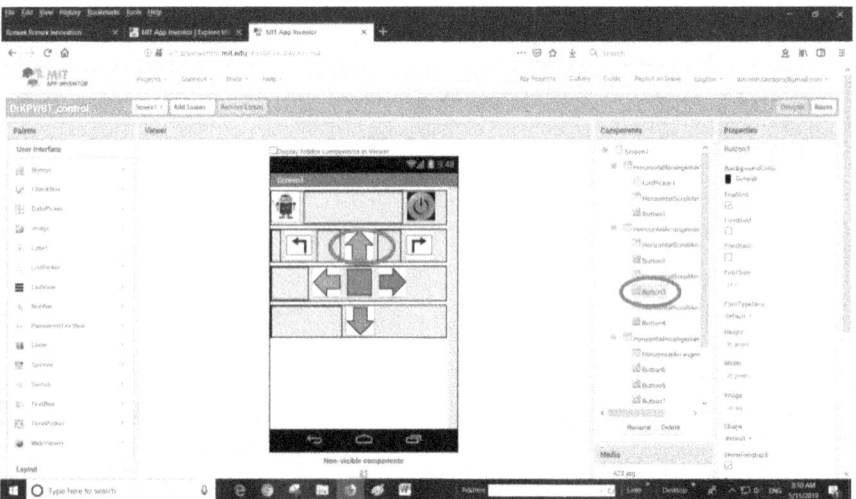

All the remaining buttons have similar functions to Button 2. Thus, we can use copy and place. Right click and choose "Dupicate".

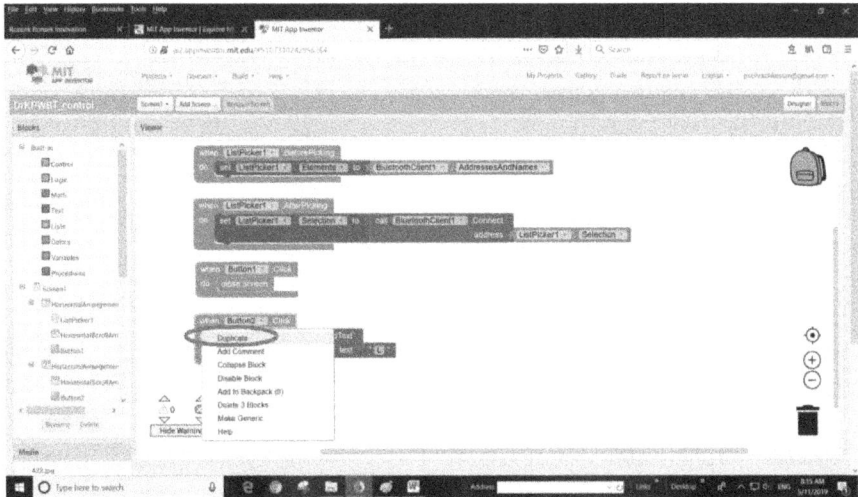

Thus, there are two similar commands. Therefore, we need to change the Button number to Button 3 and text to "u" as "UP" and move forward.

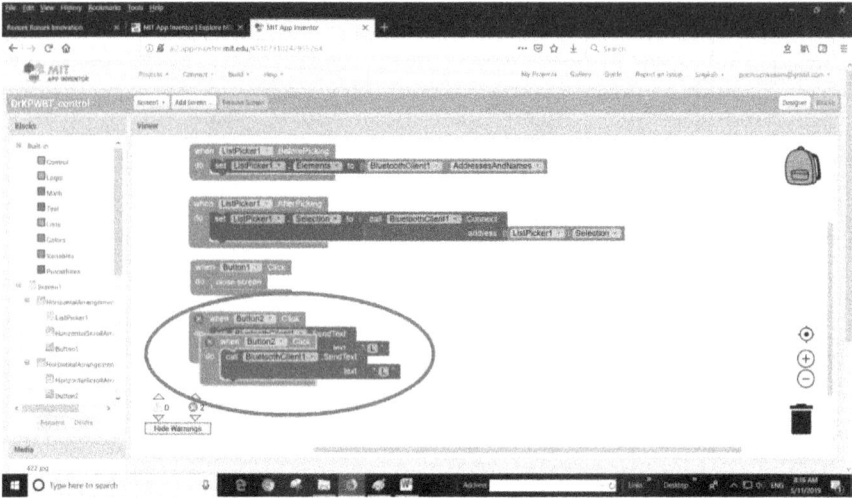

At the drop down option, choose "Button 3" and text to "u".

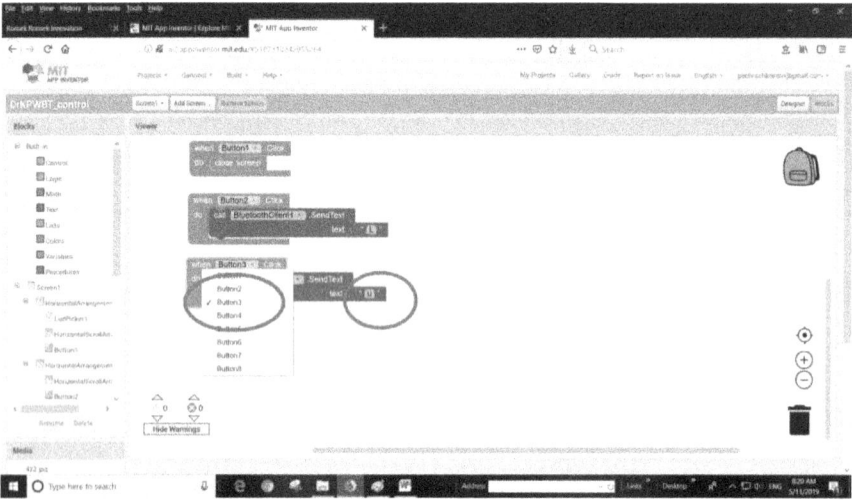

Now we have finished the Button 3.

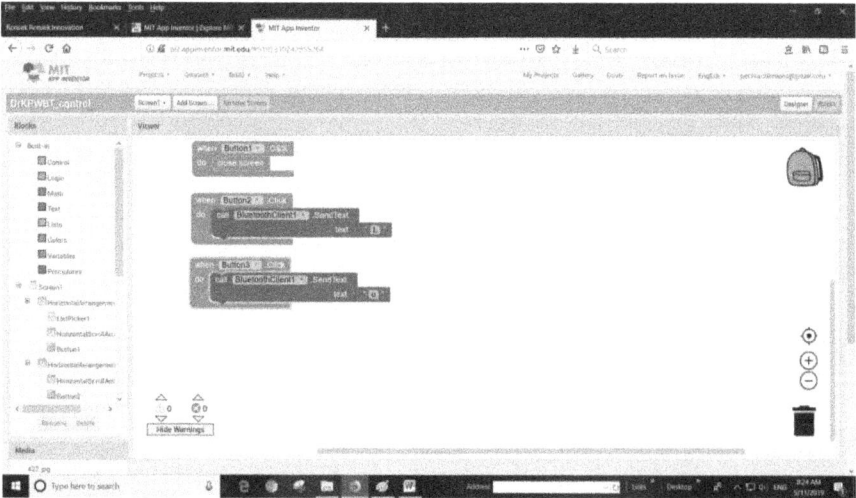

C. Right Circular clockwise Button

Button 4 is in the circle clockwise in the right direction.

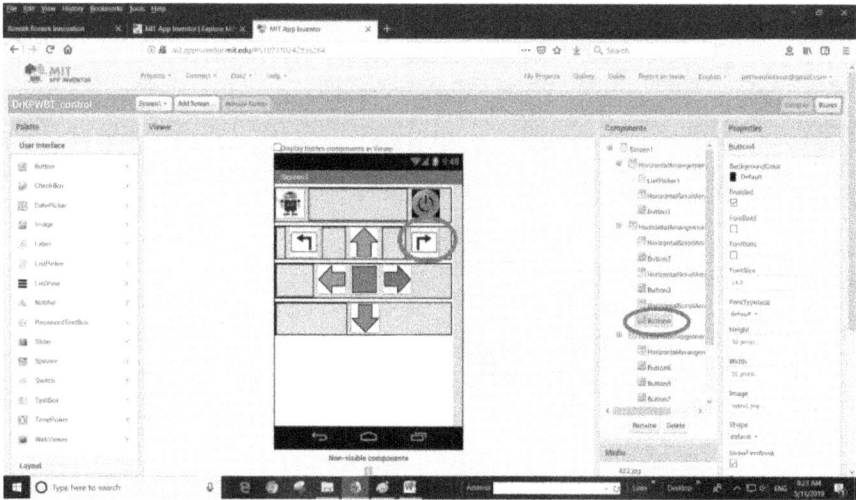

We can use the copy and place, Control key with C and Control key with V.

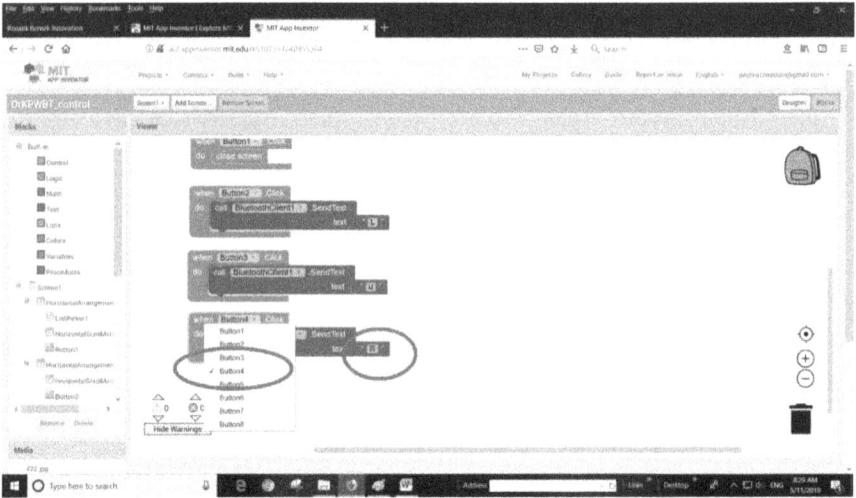

Change to Button 4 and the text to "R".

D. Stop Button

Button 5 is for the Stop Button.

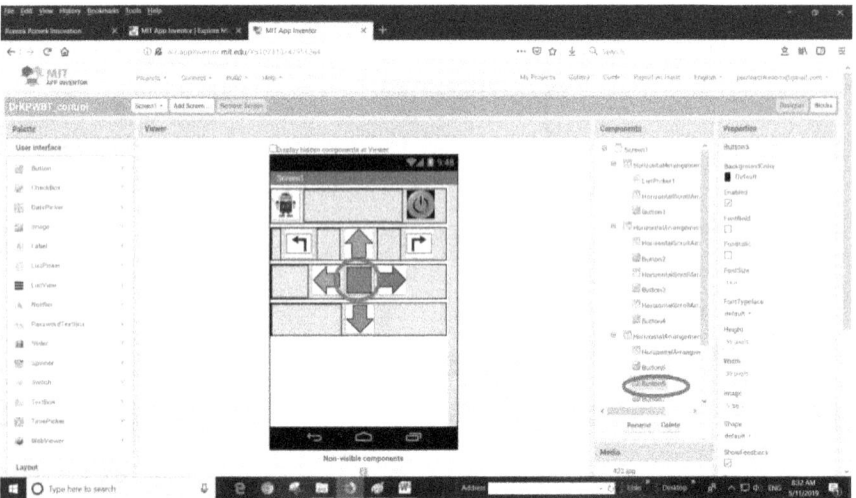

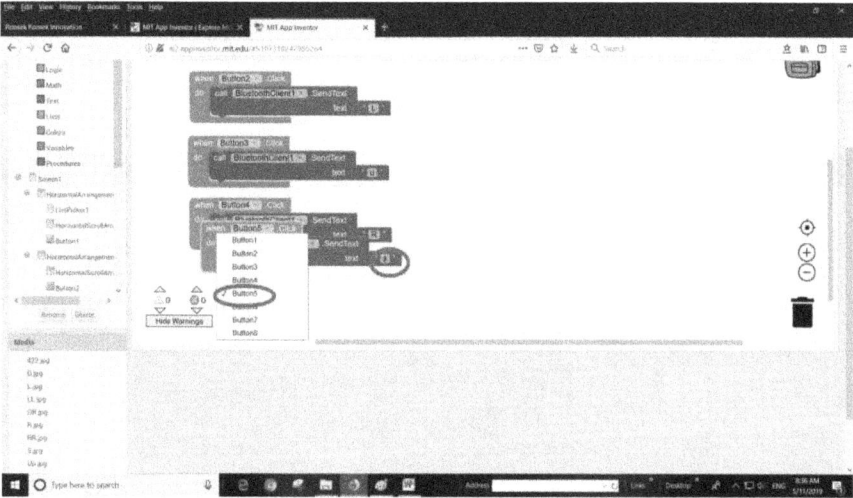

Change to Button 5 and text to "s".

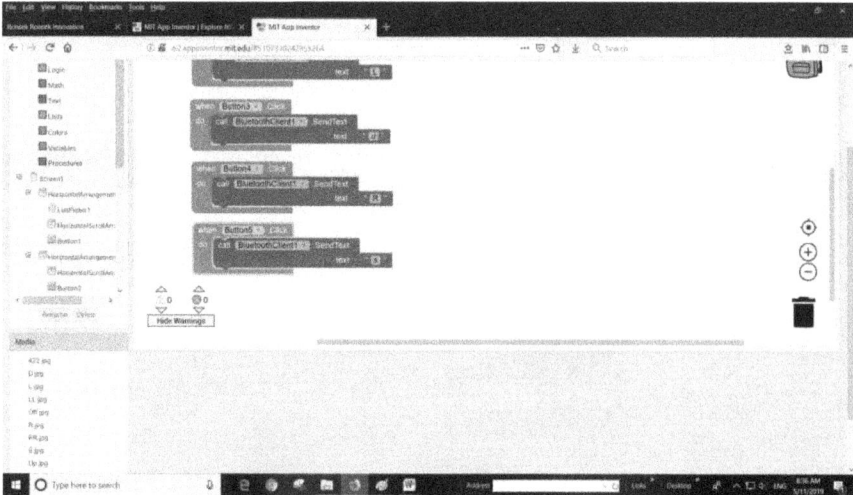

E. Turn Left Button

Button 6 is the function to turn left.

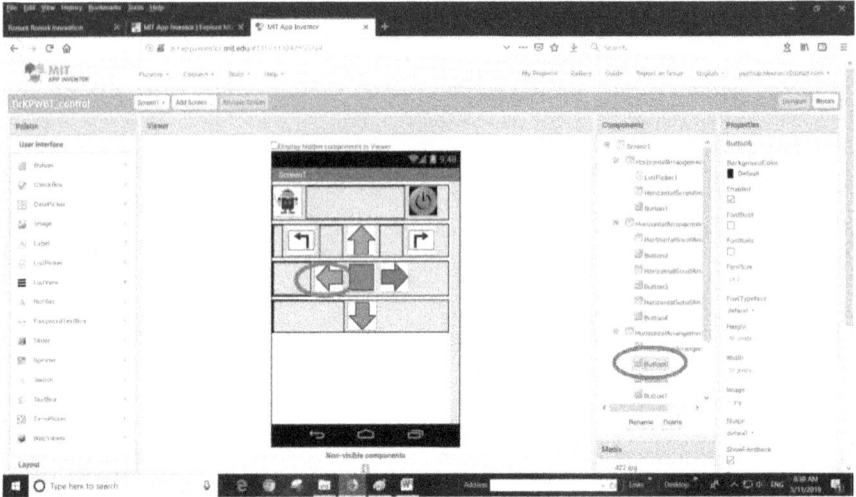

Change to Button 6 and text to "l".

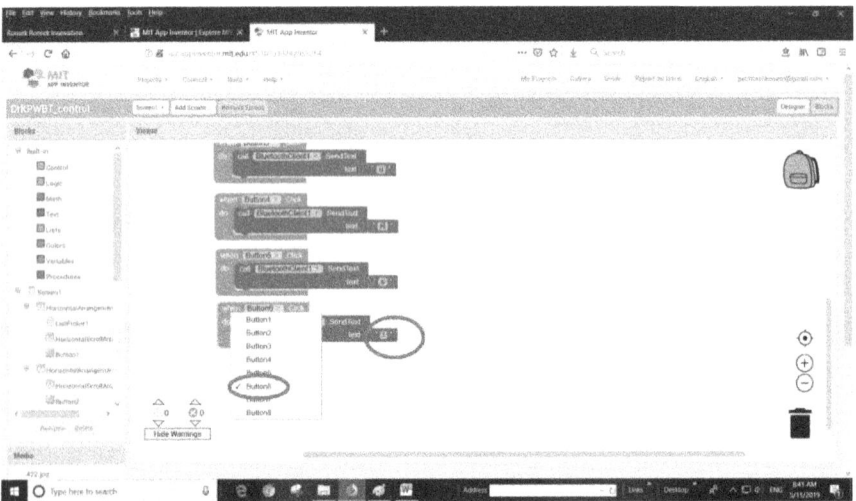

Arduino App Bluetooth Robotics

F. Turn Right Button

Button 7 is the function to turn right.

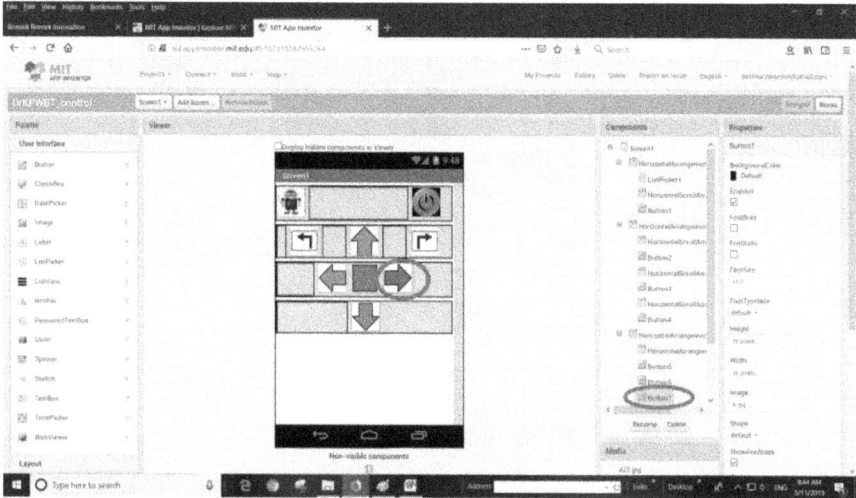

Change to Button 7 and text to "r".

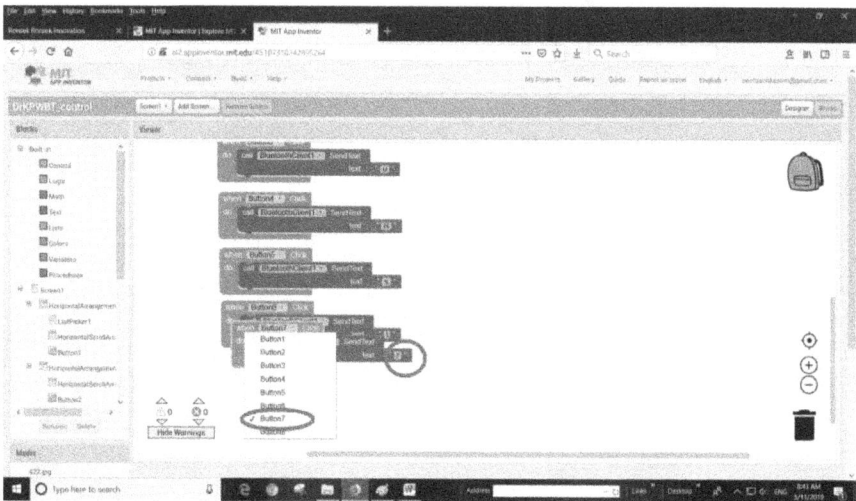

G. Down Button

Button 8 is the function to go down or backward.

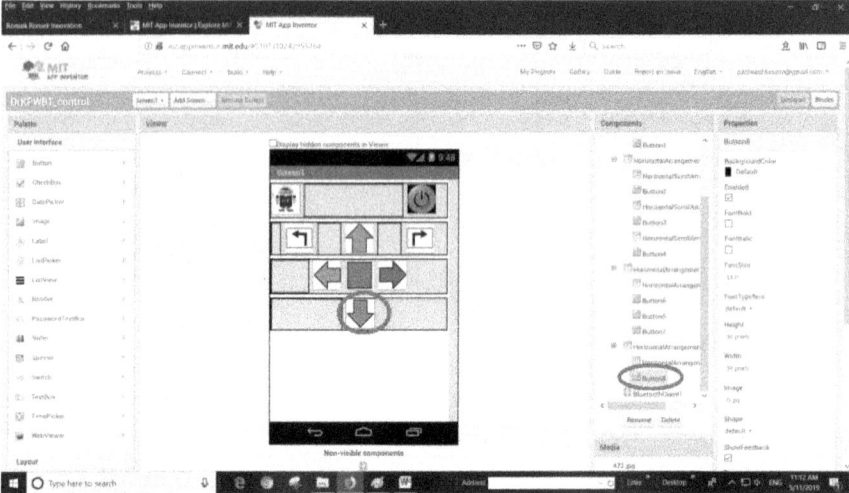

After copy the previous block, we change the Button from 7 to 8 and the text from "r" to "d"

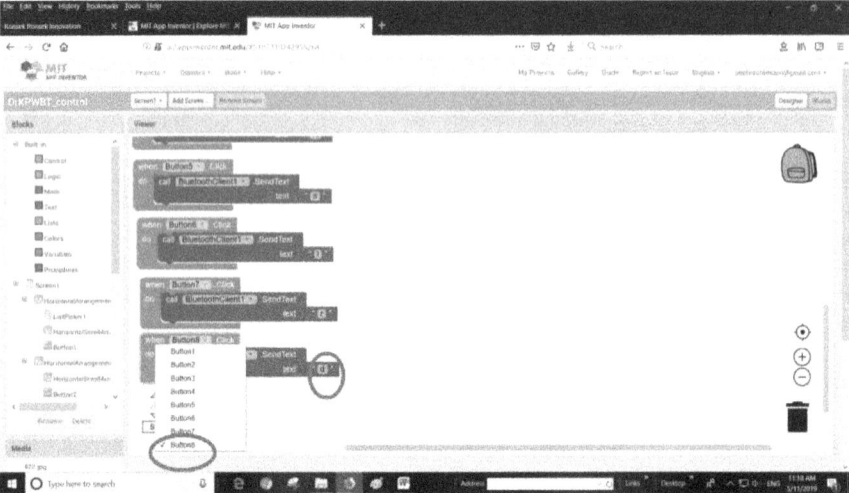

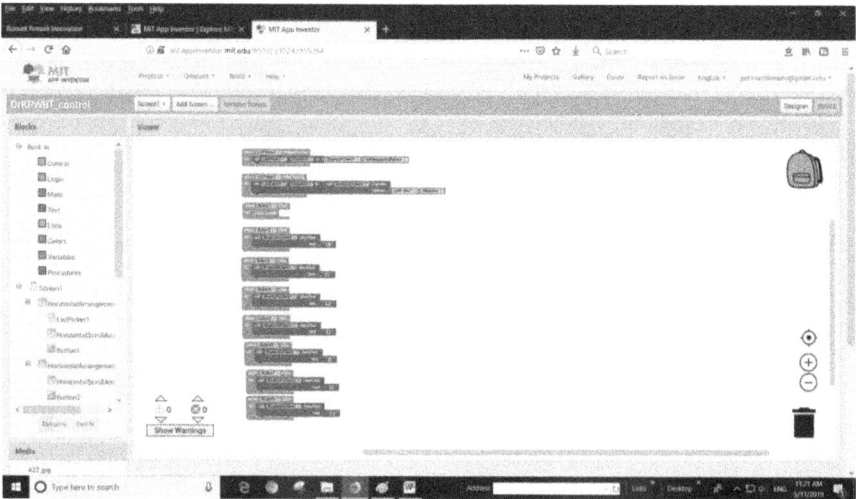

Finally, we have finished all the codes.

Next, we are going to save a file and load it on the mobile phone for later use.

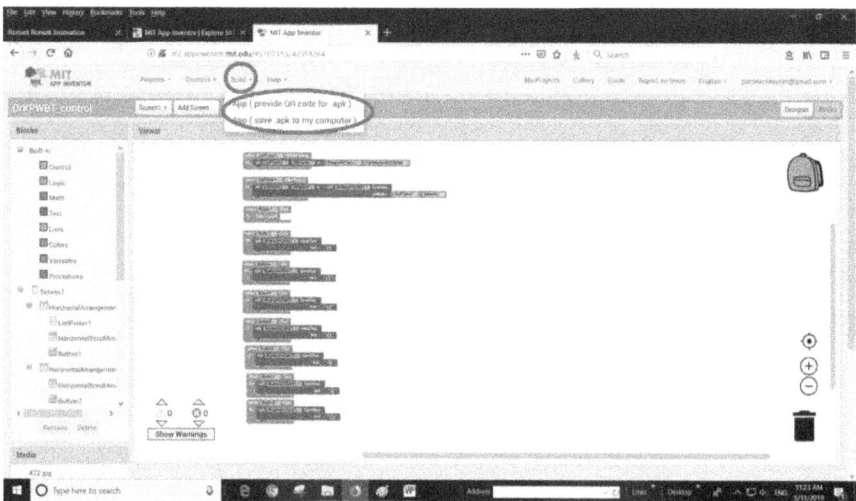

Click "Build"on the top bar, there are two options to save the file.apk to my computer or provide QR code.

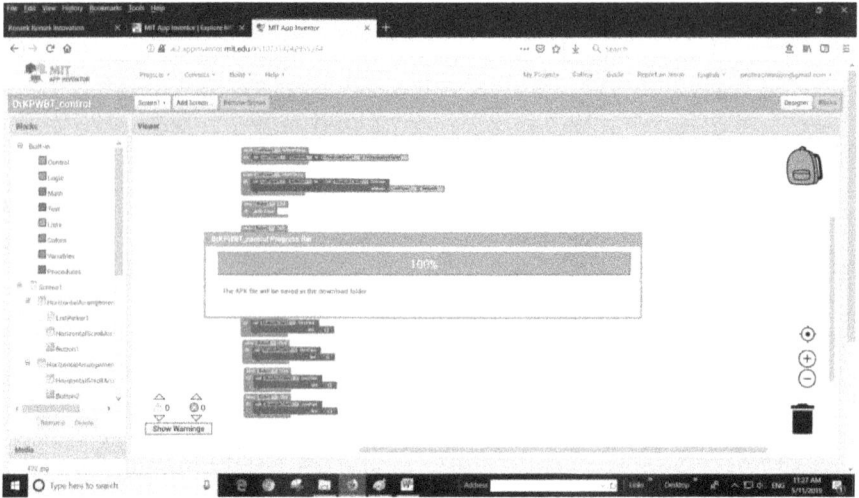

We choose to save file to computer.

CHAPTER 4

Design Circuit and Test

A. Design circuit

We start on the webpage www.tinkercad.com, which we need to join first before we can use it.

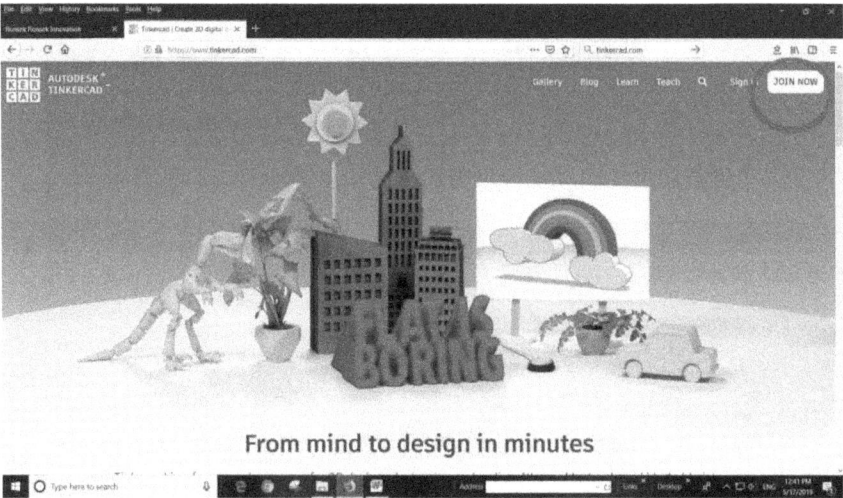

From mind to design in minutes

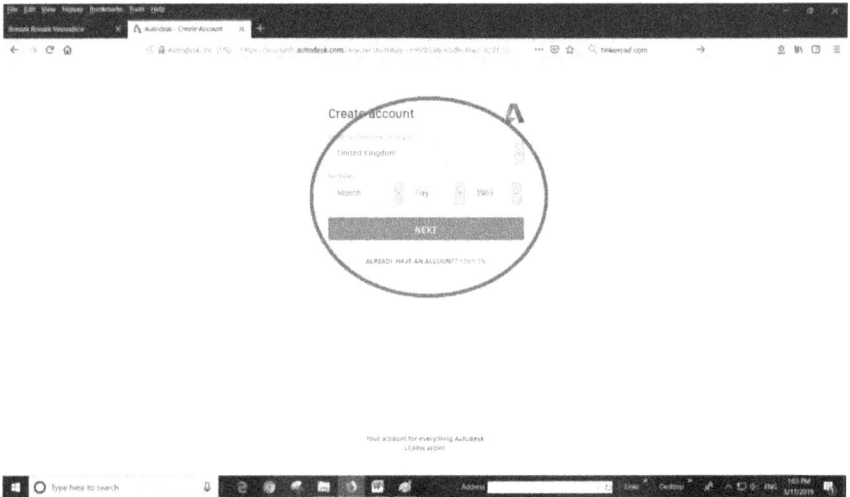

After register and log in, we will see the window below. Then, click "circuit" and "create new circuit".

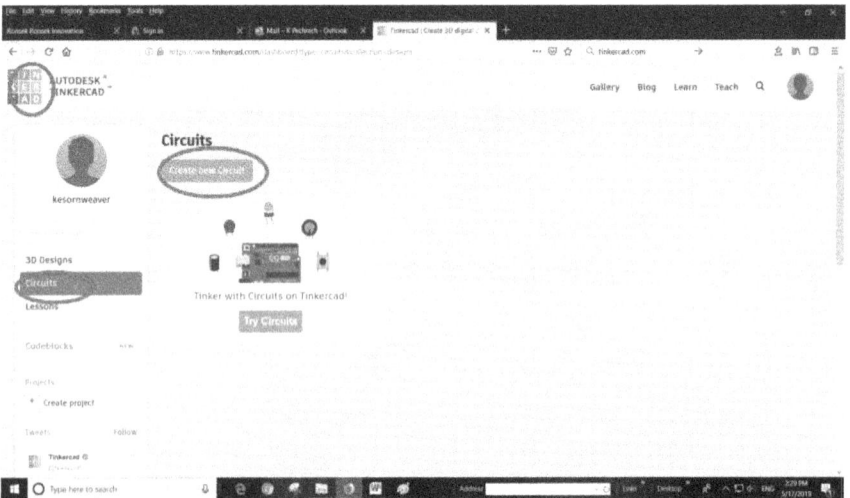

Next, at the Basic Components, choose Breadboard, Click and place in the space.

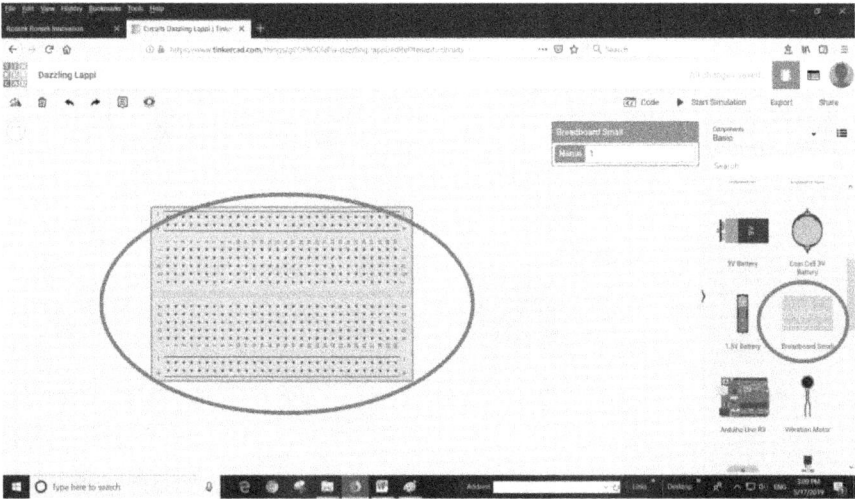

Next, we need Arduino UNO R3, choose from the
component and drop next to the Breadboard.

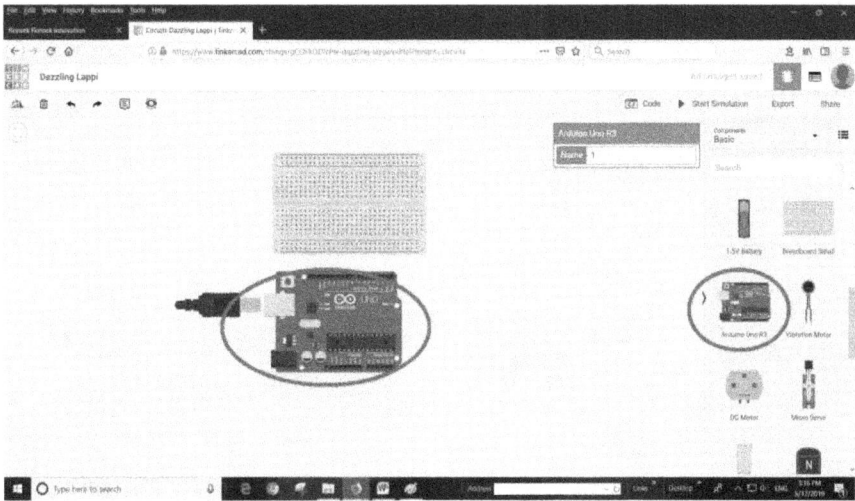

Next, we need two Gearmotors, put them by the breadboard.

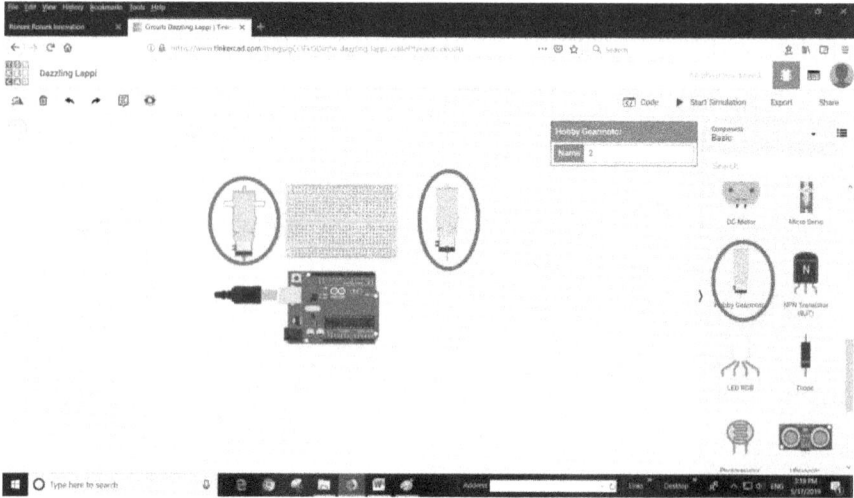

Next, we need motor drive, Integrated circuit, IC L293D, puts on the breadboard. The "White" dot is the start of the Pin number 1.

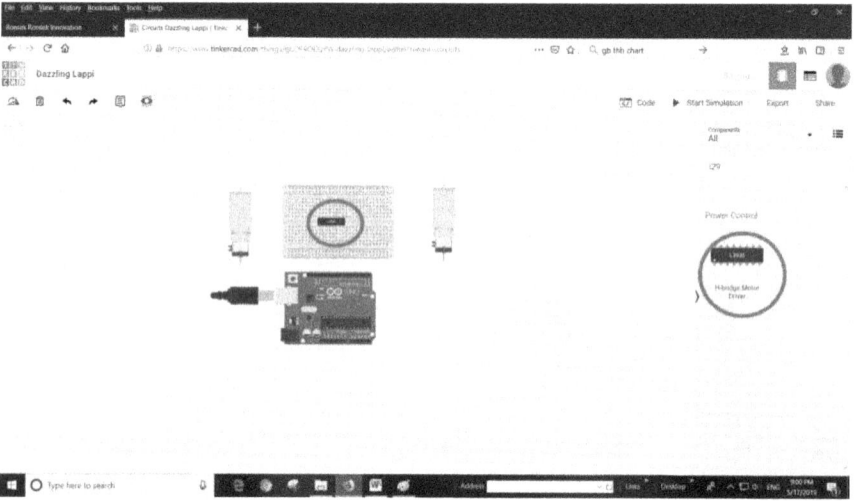

Next, we need a power supply. Type "power supply" in the search box.

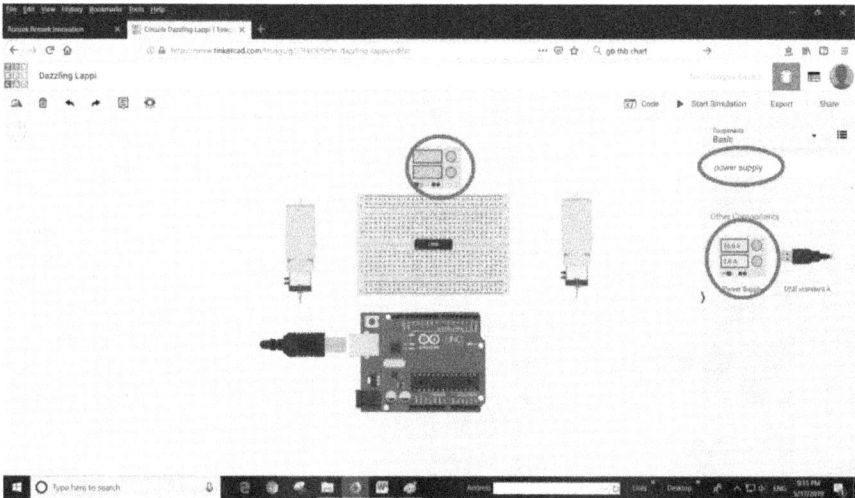

We are going to start to connect the wire as follows. Click Pin 1, 8 and 16 connects to the Red, positive.

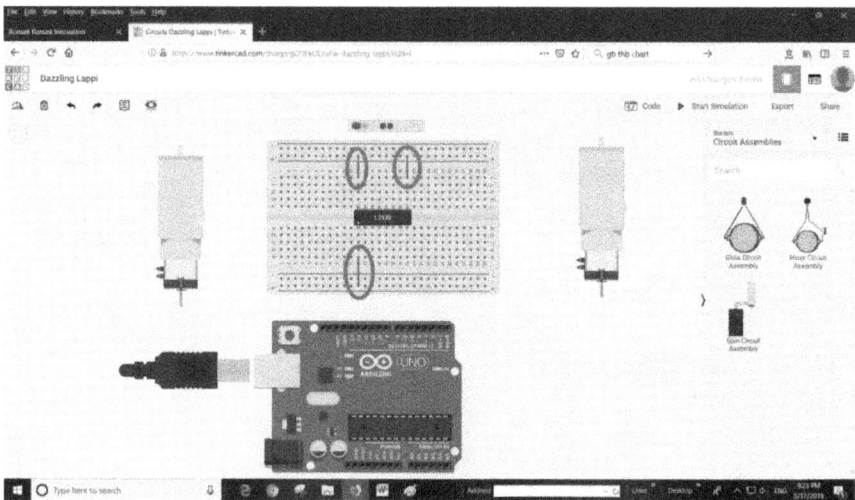

When we move cursor to the IC pin, it will show what Pin would connect.

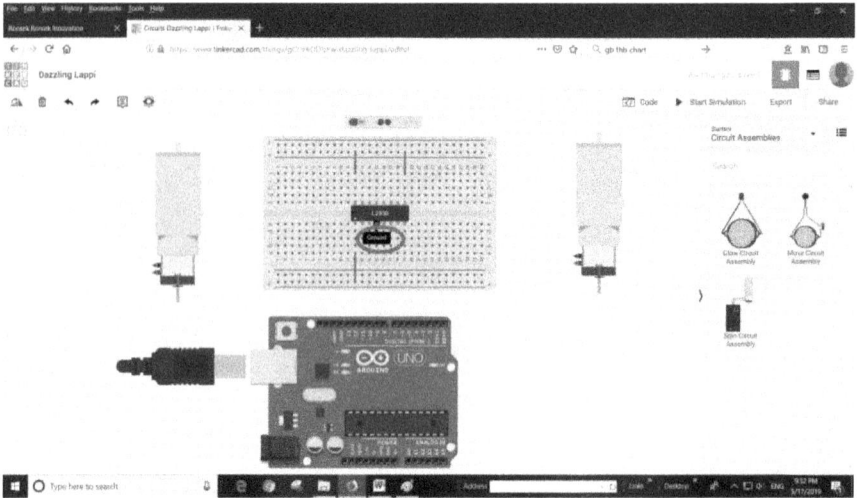

The black wire connects "Ground" from the IC to Black on the breadboard.

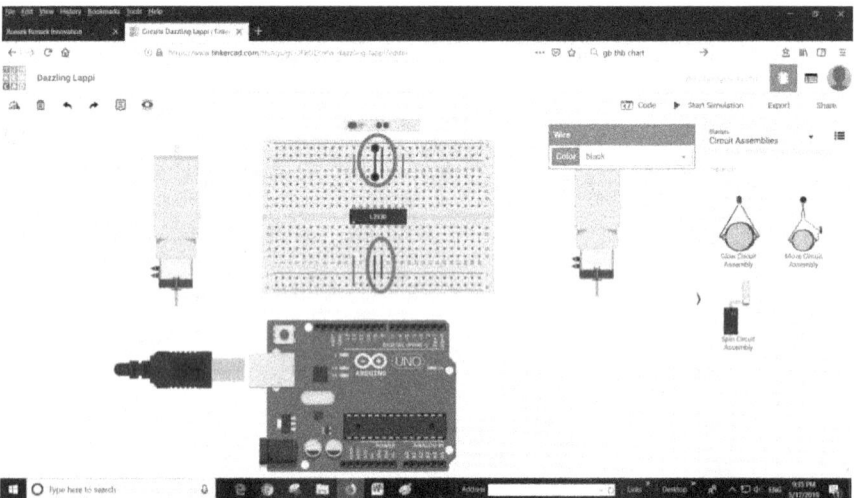

Next, we connect "Ground" from the power supply and "Ground" from the Arduino to "Ground" on the Breadboard.

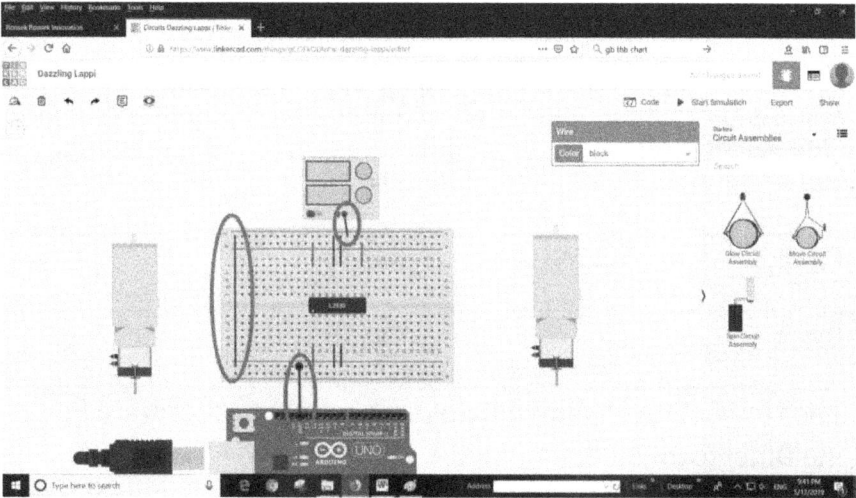

Next, connect power supply 9 Volts for feeding the motors, connect to motor drive Pin 8.

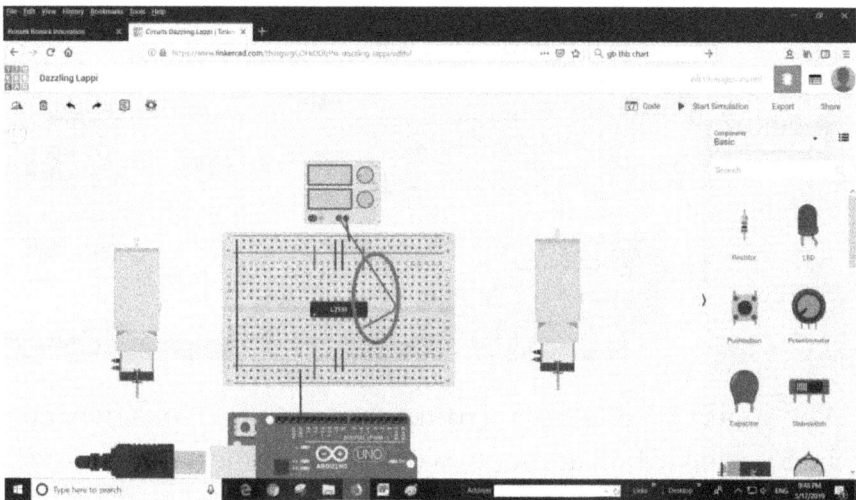

Then, we connect the motor control drive IC L293D to the Arduino as follows:

Arduino UNO	L293D Control Motor
Pin 2	Pin 2
Pin 3	Pin 7
Pin 4	Pin 10
Pin 5	Pin 15

The connection is shown below. These connections are for the Bluetooth.

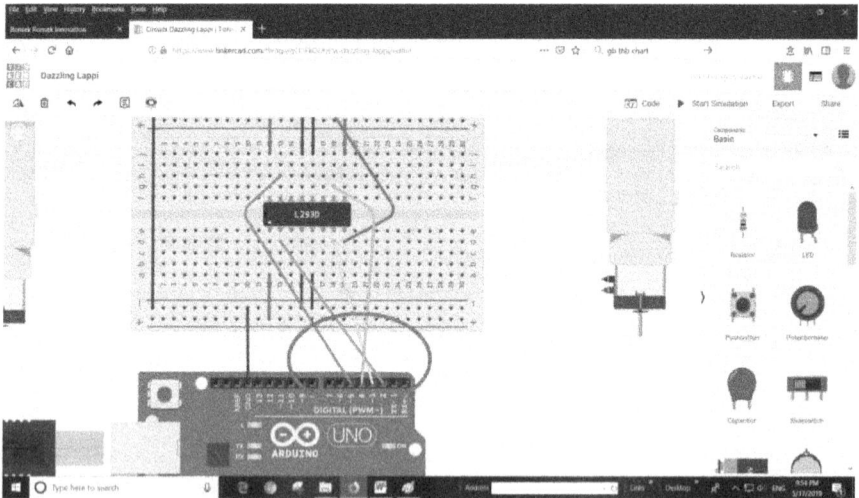

We connect 5 Volts from Arduino board to Red, positive, on the breadboard. In addition, we need to connect Red to Red on the breadboard to link the 5V positive.

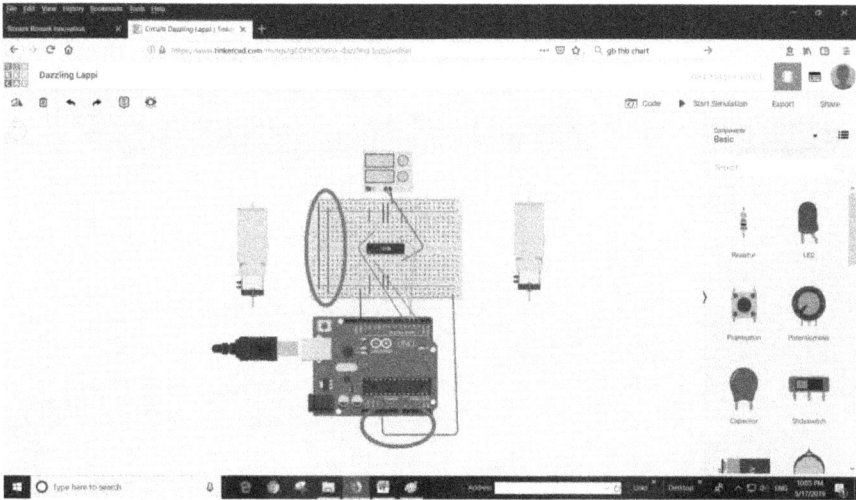

Next, we connect the left motor. The positive Red connect to IC L293D Pin 3 and negative Black connect to Pin 6.

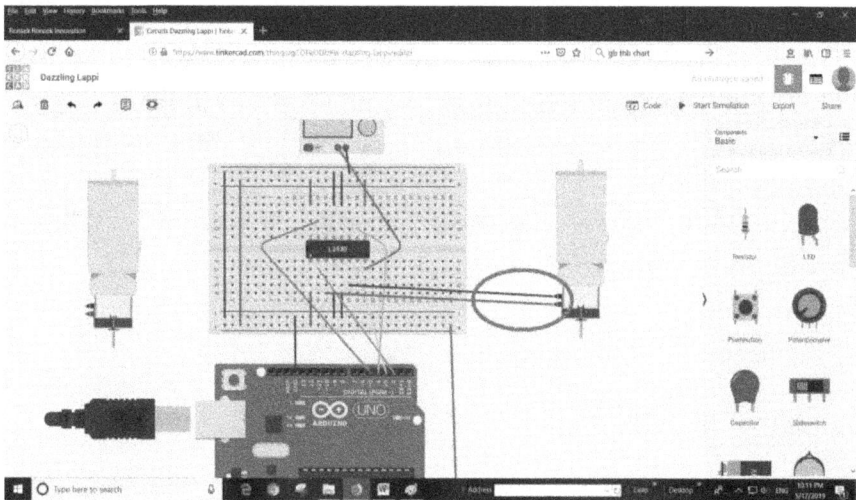

When the signal from Arduino Pin D2 is a logic "1", the output 1 of the L293D Pin 3 is positive and output 2 of the L293D Pin 6 is negative. The logic "1" is 5 Volts and "0" is 0 Volts.

However, when the logic is "0", the output 1 of the L293D Pin 3 is negative and output 2 of the L293D Pin 6 is positive.

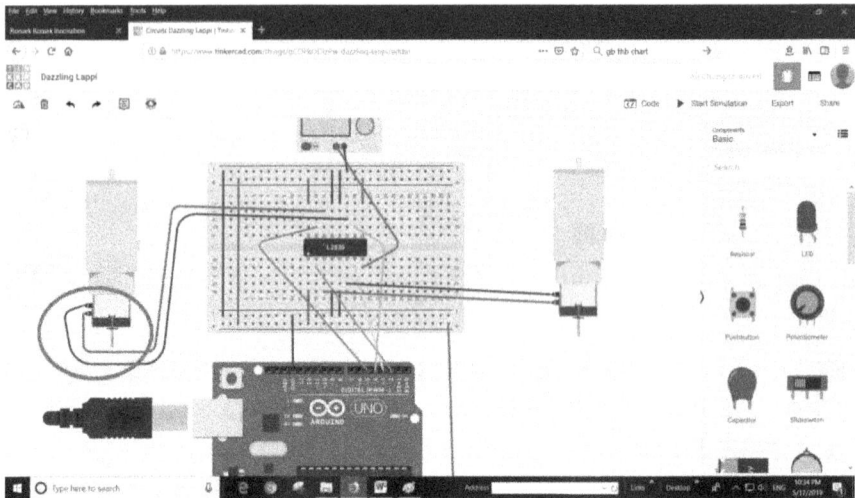

The right motor, Red positive connects to IC 293D Pin 14 and 11. The working function is the same. The signal from Arduino would come from Arduino Pin D5.

Now the wire connectors are completed. Next, we need to program it.

CHAPTER 5

Program Circuit

Create code circuit

We still use the tools on the webpage www.tinkercad.com, After we have all the wire connectors are completed. Next, we need to program it.

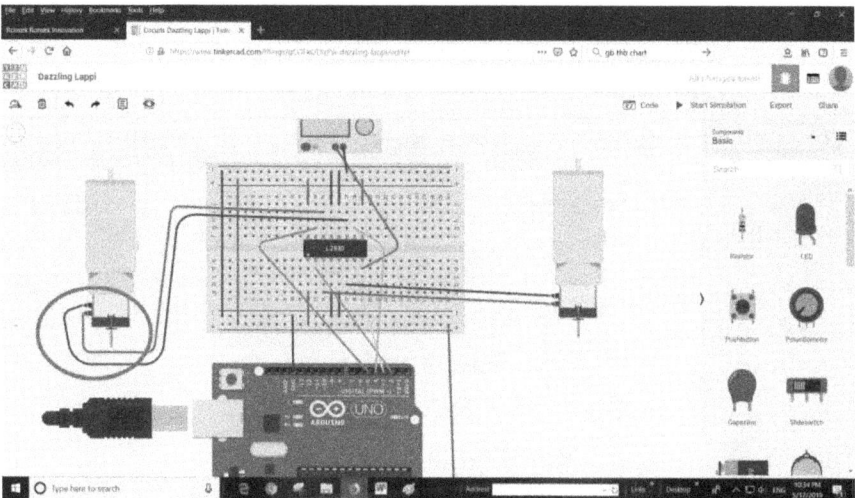

To code our circuit as shown in the next window. Click at the "Code" and there are drop down option to choose Blocks", Block+Text" or "text".

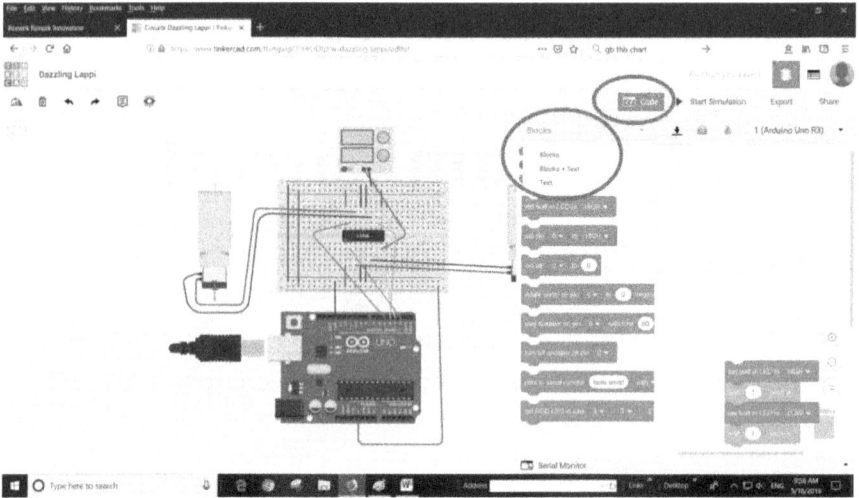

When we choose Block, the window would show as follows:

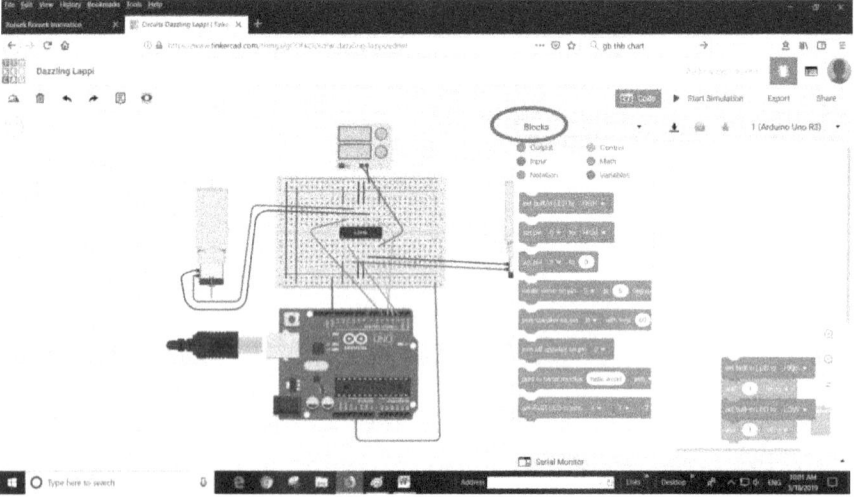

When we choose the Block+Text, the window would show as below.

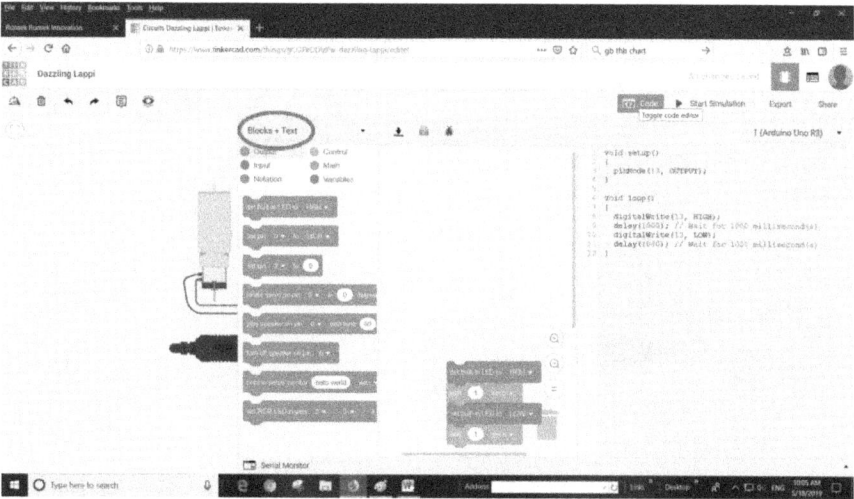

Finally, when we choose the third option, text only. There is a request to confirm before process. Any blocks you have been created would be lost.

In this case, we have not done any blocks yet. Thus, we choose "Continue".

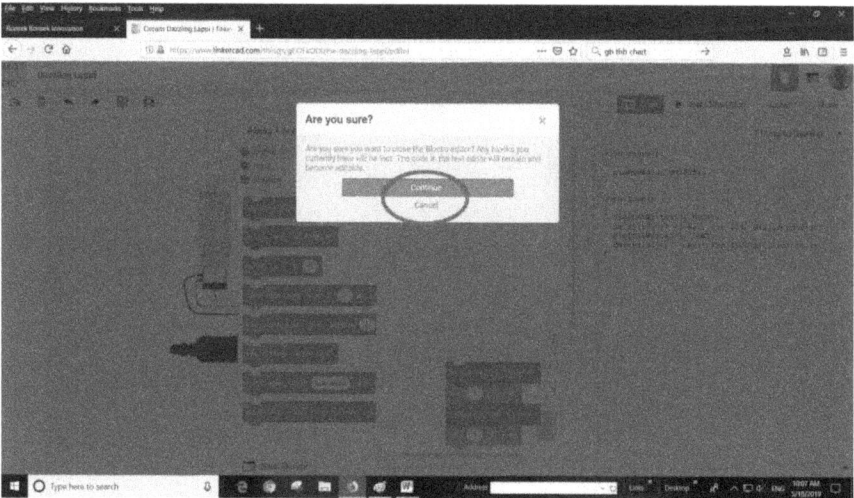

There would be only the text part show on the window.

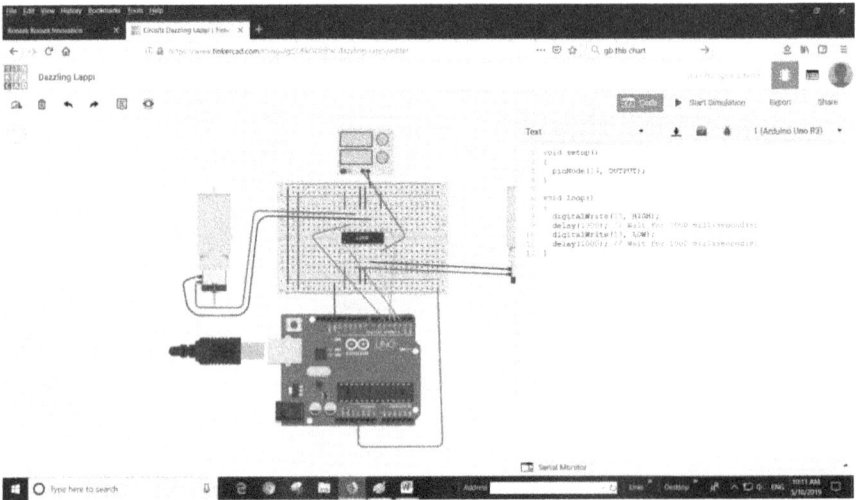

Now we are going to code the circuit. The first thing, we need to do is to define the Pin for those two motors.

We call the right motor is A1 and the left motor is B1, each of them has two wires out, which black and red. Thus, we define:

Motor A1 pin 2

Motor A1 pin 3

Motor B1 pin 4

Motor B1 pin 5

Therefore, we would use the function

#define MotorA1 2

#define MotorA1 3

#define MotorB1 4

#define MotorB1 5

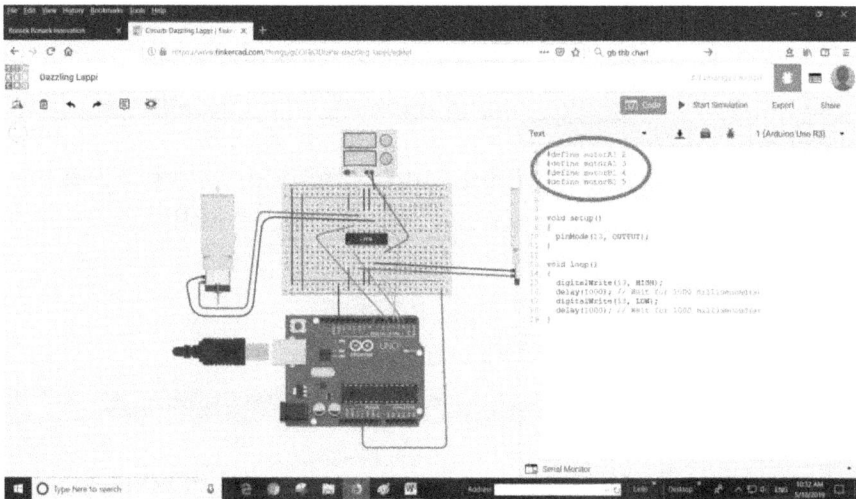

Next, we need to set up those motors and their pins. We use pinMode:

pinMode (motor A1, OUTPUT)

pinMode (motor A2, OUTPUT)

pinMode (motor B1, OUTPUT)

pinMode (motor B2, OUTPUT)

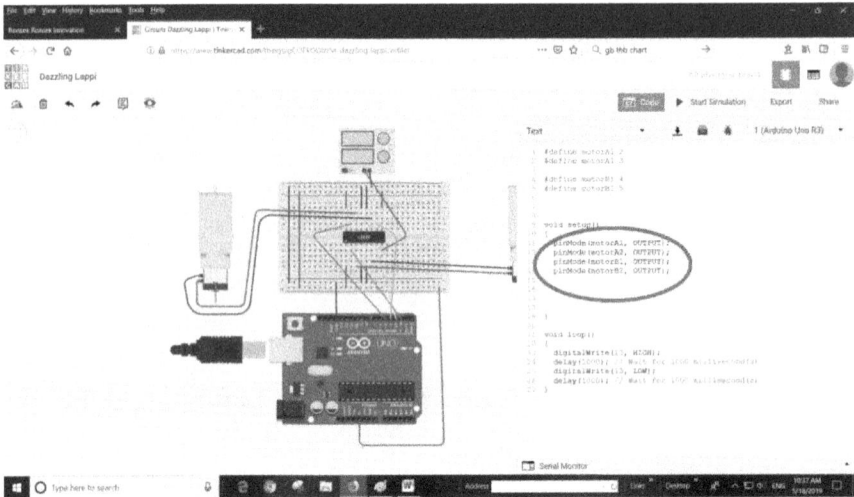

Next, we set up the communication method, which we use Bluetooth. Thus, we use Serial and the speed of Bluetooth is 9600.

Serial.begin(9600)

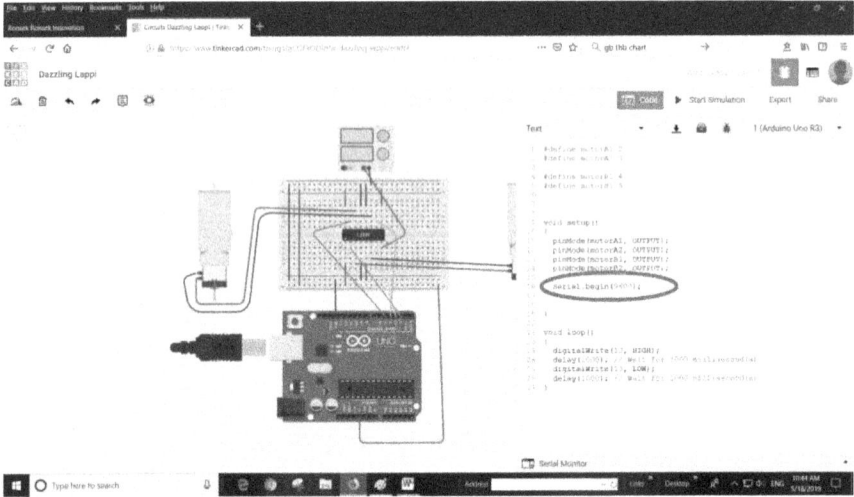

Next, we move to the Loop. We would start from the Bluetooth read the value and check if the what is that value. Thus, we have to define the value as integer, we use "val"

Int val

And in the Loop, to check what the value from the Bluetooth. If the value is "u" , both motors would move forward. The "u" is the direction that we set on the App.

The transportation will move forward if both the motors move forward in the same direction. Thus, the Red is positive and the Black is negative.

val=Serial.read()

if (val =='u')

digitalWrite(motorA1, HIGH)

digitalWrite(motorA2, LOW)

digitalWrite(motorB1, HIGH)

digitalWrite(motorB2, LOW)

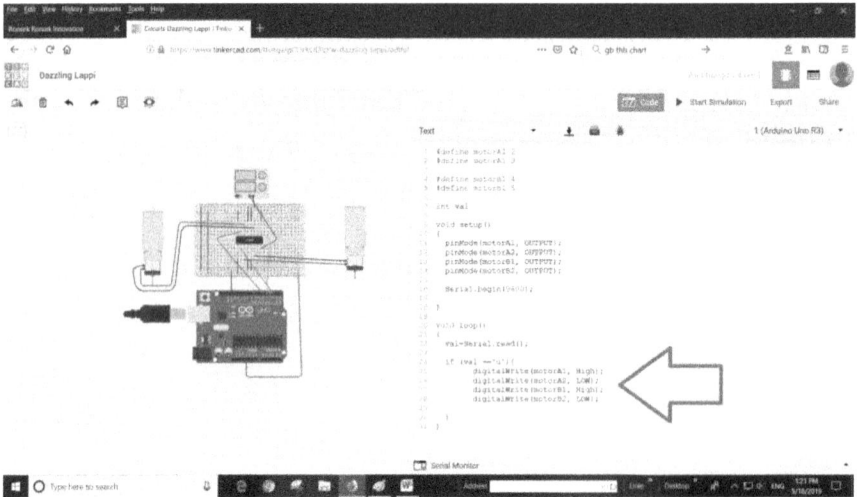

If the value is "d", both motors would move backward. The "d" is the backward direction that we set on the App.

For the transportation moves backward, both the motors have to move backward in the same direction. We could do by switching the power supply polarity. Thus, the Red is negative and the Black is positive. We use else if :

else if (val =='d')

digitalWrite(motorA2, HIGH)

digitalWrite(motorA1, LOW)

digitalWrite(motorB2, HIGH)

digitalWrite(motorB1, LOW)

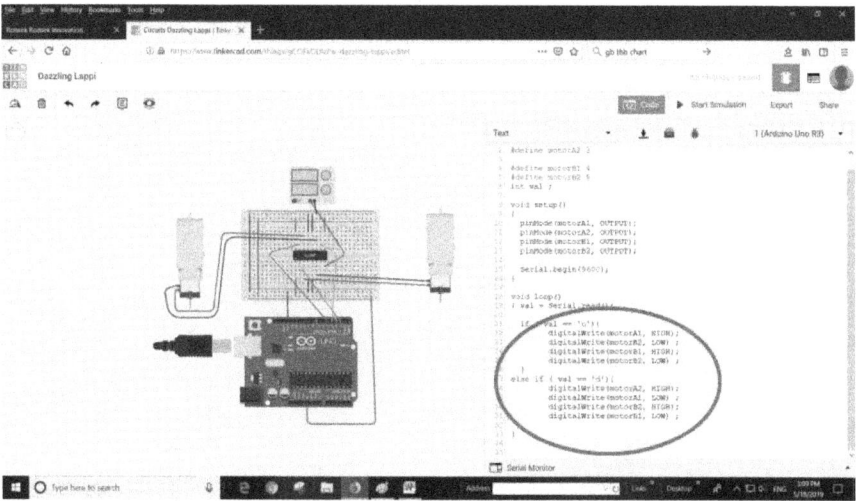

To check if the function run correctly, we can run circuit simulation.

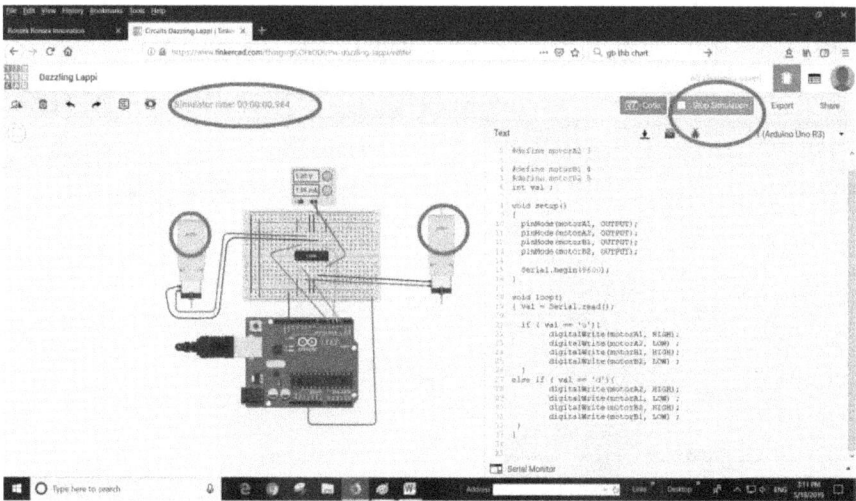

On the Serial Monitor, we type "u" and "d" to see if the motors move forward or backward as we expect.

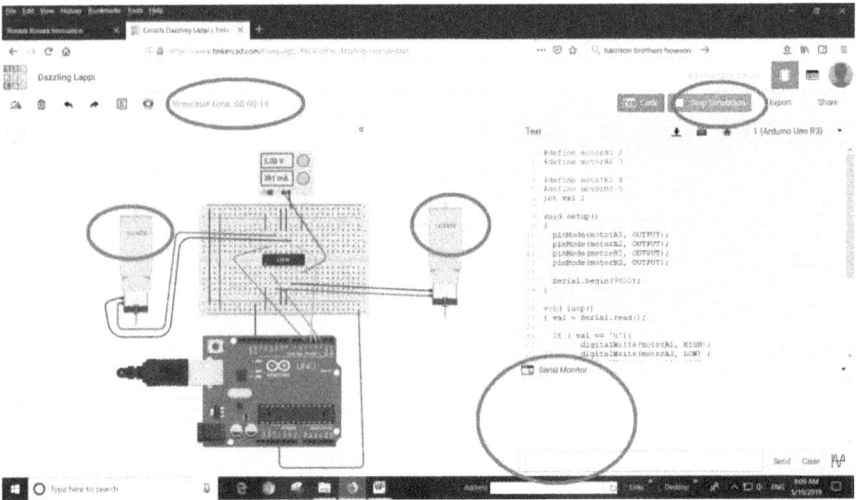

When we input "u", the right motor A1 runs at 143 RPM, while the left motor B1 runs at -143 RPM. The robot would not move forward in this case. Thus, we need to switch the left motor, Red to Pin 11 and Black to Pin 14 in the L293D.

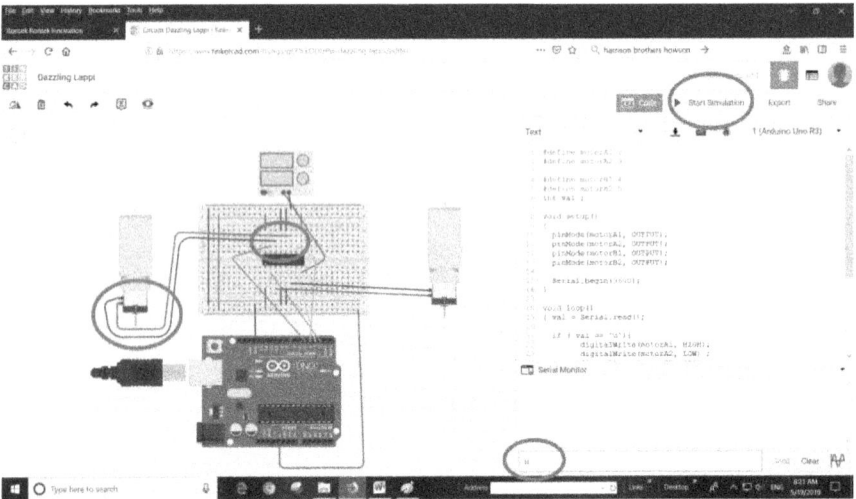

Now both motors are moving forward in the same direction when we put "u".

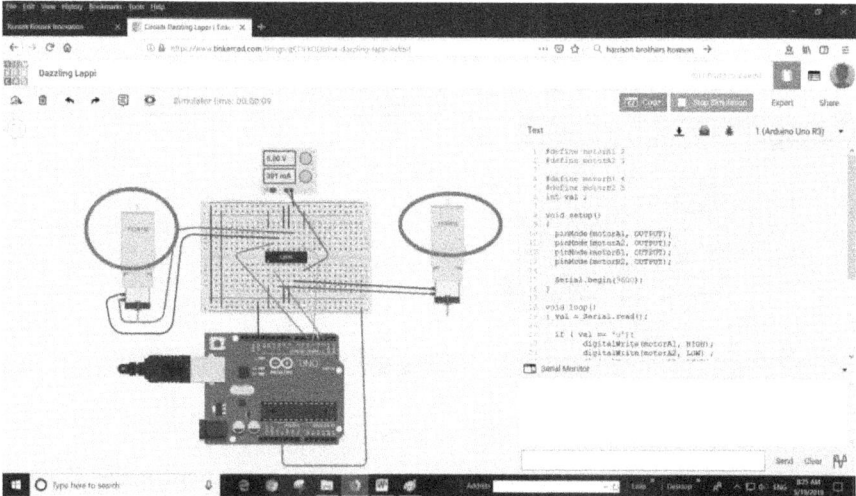

Next, we put "d" in the Serial monitor.

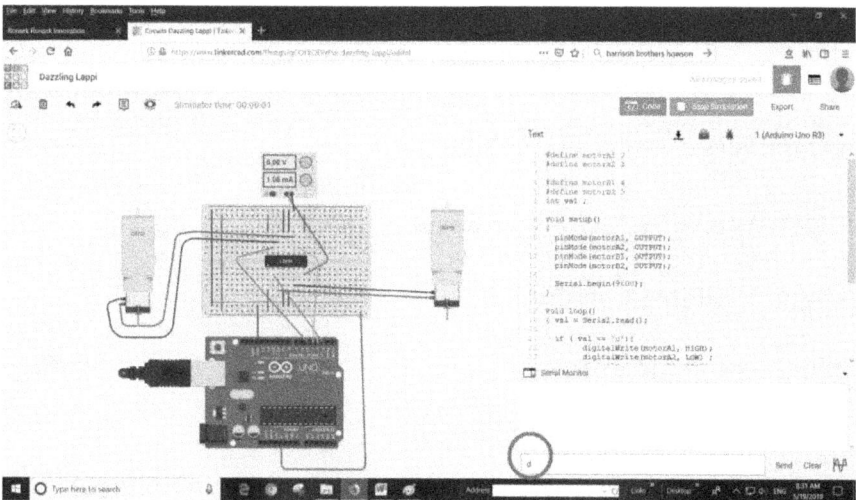

Both motors move backward with the speed -143 RPM.

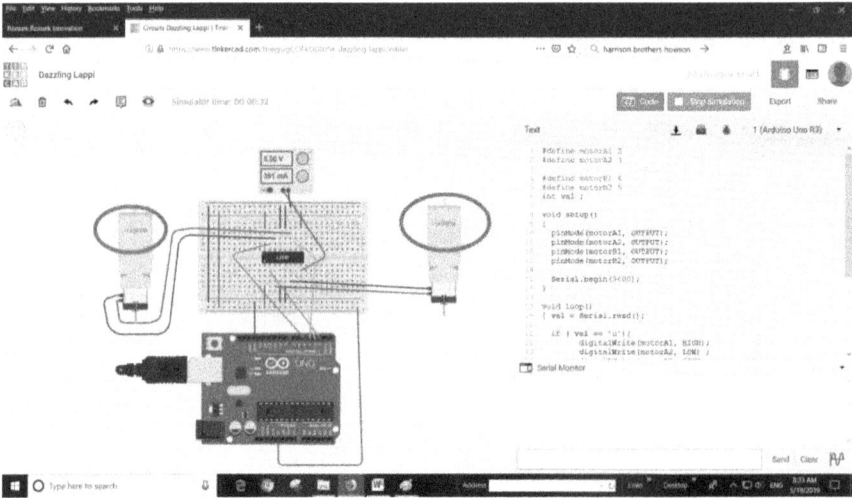

Next, we program turns right, use the same "else if". To make the robot turn right. Thus, only the left motor B1 works, while the right motor A1 stop.

else if (val =='r')

digitalWrite(motorA1, LOW)

digitalWrite(motorA2, LOW)

digitalWrite(motorB1, HIGH)

digitalWrite(motorB2, LOW)

The simulation shows that the right motor B1 runs at 0 RPM, while the motor A1 runs at 143 RPM.

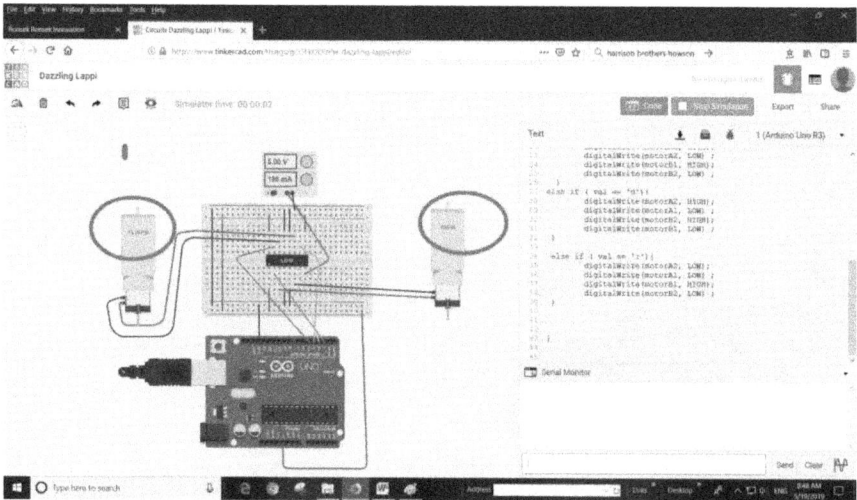

To make the robot turn left. Thus, only the right motor A1 works, while the left motor B1 not working. We program use the same "else if".

else if (val =='1')

digitalWrite(motorA1, HIGH)

digitalWrite(motorA2, LOW)

digitalWrite(motorB1, LOW)

digitalWrite(motorB2, LOW)

The simulation result shows as below.

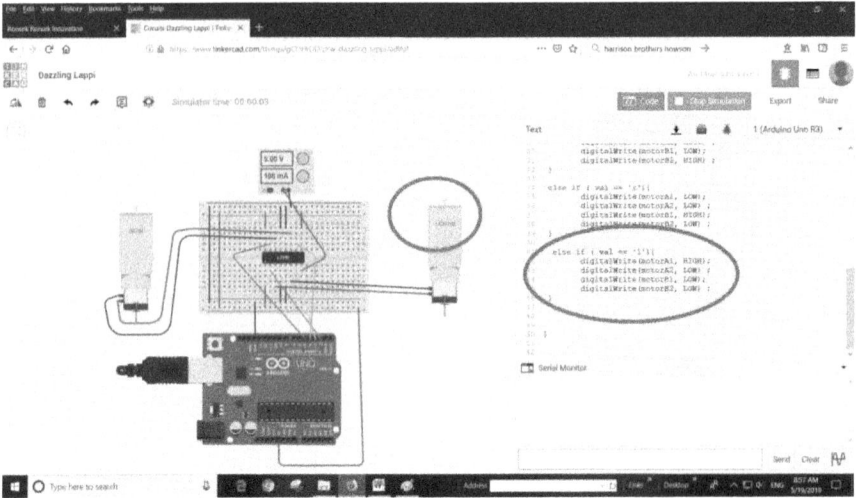

Next, the robot moves circularly in the clockwise direction. The left motor B1 has to move forward, while the right motor A1 has to move backward. Thus, for left motor B1, the Red connect to the positive(HIGH), Black connect to the negative. While, for the right motor A1, it needs to reverse. Therefore, the Red connect to negative (LOW),and Black connects to positive (HIGH). The program would be:

else if (val =='R')

digitalWrite(motorA1, LOW)

digitalWrite(motorA2, HIGH)

digitalWrite(motorB1, HIGH)

digitalWrite(motorB2, LOW)

The simulation shows that the left motor B1 move forward with speed 143 RPM, while the motor A1 moves backward with speed -143 RPM.

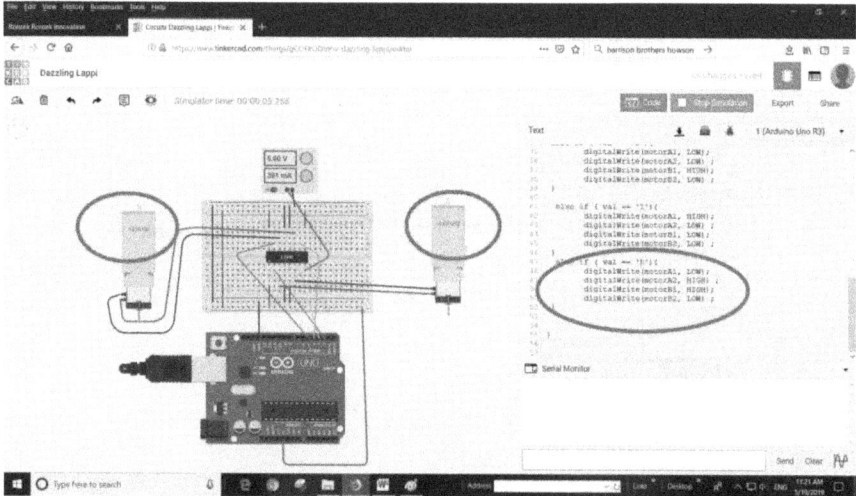

Next, the robot moves circularly in the anti-clockwise direction. The left motor B1 has to move backward, while the right motor A1 has to move forward. Thus, for left motor B1, the Red connect to the negative(LOW), Black connect to the positive (HIGH). While, for the right motor A1, it needs to reverse. Therefore, the Red connect to positive (HIGH),and Black connects to negative (LOW). The program would be:

else if (val =='L')

digitalWrite(motorA1, HIGH)

digitalWrite(motorA2, LOOW)

digitalWrite(motorB1, LOW)

digitalWrite(motorB2, HIGH)

The simulation shows that the right motor A1 move forward with speed 143 RPM, while the left motor B1 moves backward with speed -143 RPM.

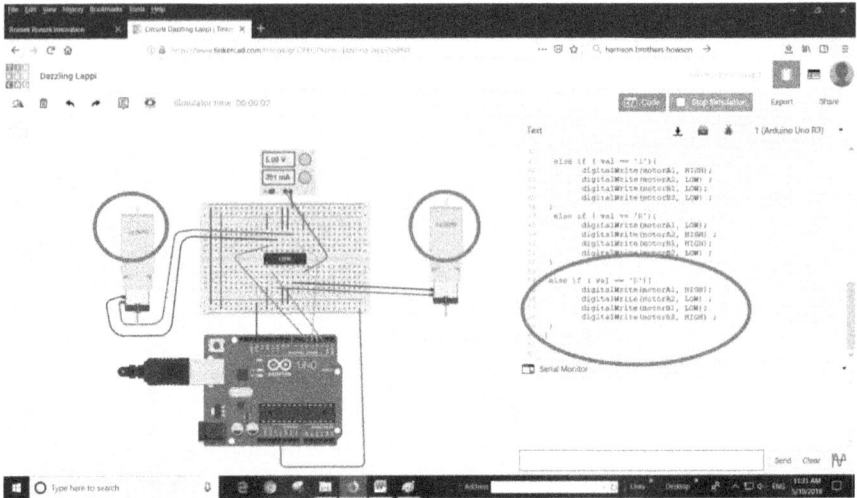

The last one is the "stop". When we choose this, every motors should stop. Both red and black would connect to negative (LOW). The program would be:

else if (val =='s')

digitalWrite(motorA1, LOW)

digitalWrite(motorA2, LOW)

digitalWrite(motorB1, LOW)

digitalWrite(motorB2, LOW)

The simulation would show 0 RPM both motors.

Then,, we need to connect Bluetooth to the Arduino board. The terminal was chosen. The first slot connects to the Red Positive on the breadboard, the second slot connects to the Black negative. The third slot, TX, connects to RX, Pin D0 on the Arduino board. The fourth slot, RX, connects to TX, Pin D1 on the Arduino board.

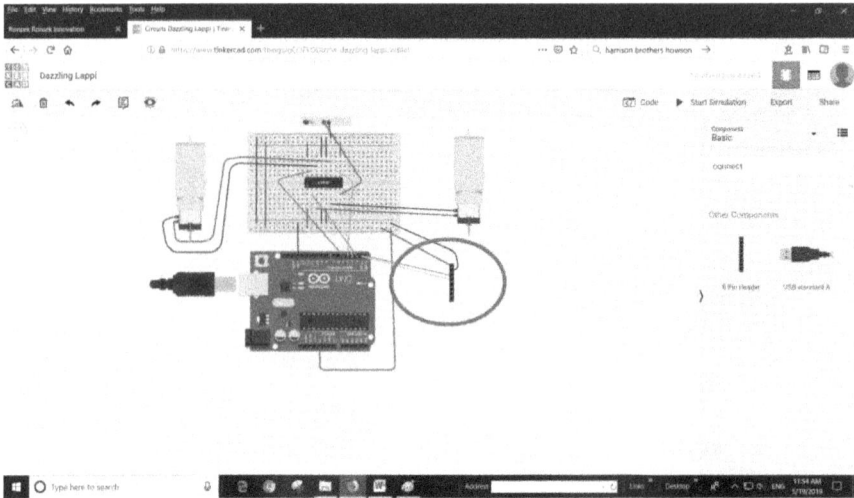

After finishing the circuit and we are happy with the running. We can download the code and upload into the Arduino for the real working robots.

REFERENCES

Kesorn P Weaver, Create Educational Robotics, Pechrach Publishing, April 2019, ISBN-13: 978-1912957-04-0.

Kesorn P Weaver, Robotics Class, Pechrach Publishing, May 2019, ISBN-13: 978-1912957-05-7

Appinventor.mit.edu

www.tinkercad.com

Arduino App Bluetooth Robotics

www.ingramcontent.com/pod-product-compliance
Lightning Source LLC
Chambersburg PA
CBHW031952190326
41519CB00007B/771